I've Been Gone
Far Too Long

I've Been Gone Far Too Long

Field Study Fiascoes and Expedition Disasters

Edited by
Monique Borgerhoff Mulder
& Wendy Logsdon

With a Preface by
John Heminway
and an Afterword by
Nigel Barley

RDR Books
Oakland, California

RDR Books
4456 Piedmont Avenue
Oakland, California 94611

First Edition

ISBN: 1-57143-054-7
Library of Congress Catalog Card Number: 96-70092

Cover Photograph: Eliott Smith
Cover Design: Bonnie Smetts
Book Design: Paula Morrison
Typesetting: Heather McLaren

Printed in The United States of America

Distributed in Canada by Orca Books
1030 North Park Street, Victoria, BC V8T 1C6

Distributed in Great Britain and Europe by Airlift
8 The Arena, Mollison Avenue, Enfield, Middlesex, EN3, 7NJ

Distributed in Australia and New Zealand by Astam
57-61 John Street, Leichhardt, New South Wales 2040

For my father who encouraged me to travel,
and in memory of my mother who loved to laugh.

MBM

We as contributors would like to acknowledge the many people in the developing world with whom we have lived and worked, and who have added such richness to our lives. Without their hospitality, be it at the national, regional, or local level, none of us would have progressed in our chosen fields as we have. Indeed, without their warmth, welcome, assistance, and friendship, our projects would have been impossible.

As editors, we would like to thank all the contributors for their stories. This volume has had a long gestation; in fact, it has been put together in the spirit of how business is conducted in the tropics: unhurried. We thank the contributors for their patience, and apologize to those whose materials had to be dropped, because they were no longer contactable at unnamed box numbers in distant lands. Truly they've been gone far too long.

Royalty proceeds will be divided between the Wildlife Conservation Society and Cultural Survival, two charitable organizations dedicated to conservation, human rights, education, and science.

Table of Contents

Preface

by John Heminway 1

Section I: Terrible Mistakes

Kelly Stewart

The Gun 5

James Serpell

The Great Parrot Hunt 19

Monica Udvardy and Thomas Hakansson

The Stubborn Snake 31

Section II: Physical Dangers

Richard O. Bierregaard, Jr.

Bushmaster in the Bidet 43

Phyllis Lee

The Ghost in the Machine 55

Herbert H. T. Prins

An East African Survival Course 65

Pieter van den Hombergh

Wildlife in Kilgoris Hospital 77

Section III: Coping with Adversity

Lisa Halko and Marc Hauser

Paper Trail to the Rain Forest 87

A. Magdalena Hurtado

My Family, Food, and Fieldwork 103

Truman P. Young
Little Criminals 117
Tim Caro
The Hottest Data in Town 127

Section IV: Clash of Cultures
Elizabeth L. Bennett
A Trip to the Che-Wong 141
Dorothy L. Cheney
Social Anthropology at the Emali Hotel 151
Monique Borgerhoff Mulder
Gitangda Is Great 165
Kate Kopischke
Bush-League Medicine 181

Section V: Research Communities
Andrew Grieser Johns
In the Forest Without a Dog 197
Margaret Symington
Siete de Enero 211
John Symington
Jungle Love 227
Ronald E. Cole
The Birth and Death of a Very Fine Pit 235
Robin Dunbar
Innocents Abroad 251
David Bygott and Jeannette Hanby
Back Seat Observers 263

Afterword
by Nigel Barley 283
About the Authors 287

Preface

by John Heminway

For years I've called them "boffins." Dismissive, I admit. You might sight them clustered on the verandah of the Norfolk Hotel, eating what appears to be their first meal in weeks, brutish in manner, furtive in glance. They dress according to arcane codes—in khakis that may have served as both sleeping bag and napkin, in buffalo trodden bush hats and boots that appear soldered to feet. Move closer and you note they pepper their conversation with boffin-speak: "Landscape ecological mapping," "relevé," "pugilistic piloerection," "food separation neurosis." Watch out: now they're counting change (Waiter, can that smile—it won't work). And here they're heading for their car! My, oh my! Will this prewar Land Rover start? Its door is covered in babble, crediting far away institutes of higher learning, its interior dank with floppy disks and disordered camping equipment (no doubt borrowed). A heavy vapor trail soon streams from the exhaust. Metal grates upon metal above the din of Nairobi's noon-day rush. Then they vanish as boffins do, the wind mercifully at their back.

Your only proof of a positive boffin sighting will be a telltale pool of radiator water mixed with engine oil, splashed onto pavement.

That was long ago. Now, I'm happy to report the code of silence that once enveloped the lives of boffins has been broken—all thanks to a ground-breaking behavioral study entitled *I've Been Gone Far Too Long*. Not only does it deal with all variants of the species, but it achieves its goals through the magic of autobiography. This book does for boffins what Iain Douglas-Hamilton and Cynthia Moss's work did for elephants, what George Schaller accomplished for gorillas and lions.

Now, at last, we can ruminate on the most illusory aspect of boffin behavior—motivation. Why do they light out for the most uncomfortable of all earth's locales, deprive themselves of balanced diets, seek out the company of incompatible people, undergo trial by mosquito, multiple flat tires, rejection by former lovers, death and, worst of all, LOSS OF DATA—all in the interests of science?

To my surprise, I discover I have much in common with boffins (I know I'm verging on anthropomorphism), especially in discovering we share the following sensations: *shame* at being "out foxed" by animals (or, in Truman Young's case, "out hyraxed" by them), *love* in the jungle (à la John Symington), *voyeurism* of Maasai sexual mores (thanks to Dorothy Cheney), *admiration* for eccentrics (as in Kelly Stewart's for Dian Fossey) and occasional *attraction* to and *despair* of tourism (Monique Borgerhoff Mulder and Robin Dunbar). Most unexpectedly, I find that every boffin encountered in these pages is gifted in yet another way. In reading these accounts, there were moments I laughed so loud fellow airplane passengers recoiled. I now can say with certainty that boffins roaming the outer reaches of Indonesia, South America, and Africa possess *a sense of humor* (possibly unique to them) that will have a similar effect on other hominids, ourselves included.

There's more. I know there is a quality I'll never share with boffins: patience. None of this book's tales waste our time with self-importance, pomposity, pedantry. The authors of *I've Been Gone Far Too Long*—writing no doubt for themselves, almost as if each incident is a private joke—are inclined to make light of whatever abilities inclined them towards lives of scholarship. Left unspoken is the one unmeasureable distinction they all seem to possess—a mind-numbing ability to sit "it" out. Read this book and you will know with certainty that natural science researchers possess quantum overloads of quiet determination and tenaciousness. Tragedy, loss, poverty, discomfort, rejection, even bureaucracy never divert them from the goals of science.

And so, I read these accounts with unabashed admiration. This book puts a human face on data-filled careers. It makes individuals we always accepted as admirable in—dare I say?—a monochrome sort of way, now altogether dazzling. Humanity emerges from every tale like morning glories after rain. I leave each account filled not only with new insights into anthropology and the natural sciences, but with an uncontrollable desire to meet each author. I also leave, twitching with envy.

Maddeningly, I doubt I'll ever again be able to say "boffin."

Section I: Terrible Mistakes

The Gun

By Kelly Stewart

I HAD BEEN dreading it, but I knew it was bound to happen. One day, she would ask me to carry the gun into the forest. I wouldn't have minded so much had it been something really obvious and cumbersome, like a rifle or a shotgun. Somehow, that would have seemed less damning. But Dian Fossey was not interested in anything as pedestrian as a rifle. No, this was a pistol, a nasty little Beretta, the kind of gun that normally appears as a bulge under gangsters' jackets. It was the only type of firearm that Dian kept. The Beretta, the Smith & Wesson—they were the armaments in her war.

The whole scene was pretty weird, but I could understand how Dian got involved the way she did: how she slipped into her Joan of Arc outfit. Back then, conservation in the Virunga Volcanos was dormant, and things looked bleak for the mountain gorillas. It was years before they hit the big time, before all the "Gentle Giant" publicity and money-pouring campaigns. In 1973, when I first went to Rwanda, the general picture was of a dwindling gorilla population, pushed high up into the mountains by hunters' spears and humanity's clamor. Officially, the Virunga Volcanos were a National Park, but the forest was crawling with cattle and poachers. The gorillas had nowhere left to go. They were under siege, and Dian took up the torch and joined them.

She needed recruits and I bounced along straight from college graduation, all starry-eyed about the prospect of living in the forest and studying gorillas. Dian was my heroine. I had read her article in *National Geographic* about 100 times. It had not mentioned guns, but I found out soon enough about Dian's war.

It was a late afternoon during my first week in the Volcanoes, and I was returning to camp after a day of observing the gorillas in Group 4, Dian's main study group. The muddy trail had just lead out of the forest near the Zaire border and onto the big meadow. It was a long stretch of golden-yellow grass, dotted with waxy lobelias with flower spikes rising ten feet into the air. Littered about were old logs and tree stumps, upholstered in velvety green moss and the occasional pink orchid. Along the northern edge of the meadow, the steep slopes of Mount Visoke rose like a wall, covered in a tangle of vegetation, with all different shapes of leaves and shades of green. To the south was Hagenia woodland, where wispy beards of pale lichen hung from the tall spreading trees and made them look ancient. This was African forest at 10,000 feet, where the evening light was so deep that colors glowed and every blade of grass cast a shadow.

So I was just ambling along, gazing at that light and listening to the silence when suddenly I heard a mad yelling of one voice that grew louder as I approached the first cabin. There was big trouble in camp. A herd of cattle had moved through, transforming Eden into an abandoned stockyard. All around was trampled vegetation, churned up mud, and great splats of dung. The four Rwandans who made up the camp staff were standing nearby, shaking their heads at the scene and making little sympathetic noises with their tongues. Dian was stamping back and forth, sucking violently on a cigarette, and letting go with salvos of abusive Swahili interspersed with the ripest English and French. I had never seen anyone so mad. The air around her almost shimmered and her eyes looked absolutely black. Before I could speak, she stormed away up the camp trail. She was going after the stragglers.

The shots kept sounding over and over, until it was dark. I sat alone in my cabin listening to them, wondering what I had gotten myself into. Dian told me later that she shot thirty cattle that evening, but I think she was exaggerating. No more than three carcasses were ever seen by anyone else. Still, killing even three cows with a pistol couldn't have been easy, not for someone who loved animals as much as Dian did. She returned to camp exhausted and distraught and said to me, "If you want to work here, if you really care about the gorillas, you have to make a choice. There's no gray, only black and white."

I envied her resolve but frankly, I could see nothing *but* gray, a thick enveloping mist. While there was no doubt that cattle could destroy the forest, it was also true that the price of a cow—about thirty dollars—was equal to the average Rwandan's yearly income. This was a country of subsistence farmers. Could we really just come in and blast away at their livestock because we thought gorillas were terrific?

Alongside the ethics of cattle-killing was the practical and more immediate question, *"Can't we get in trouble for this; serious trouble, rotting-in-a-Rwandan-jail-type trouble?"* Back then, all foreigners at camp, including Dian, worked in Rwanda on tourist visas. I was sure that we did not have permission to shoot peoples' cattle, even though they were in the Park illegally. I remembered the wooden-faced military police I had seen in Kigali with their stiff green uniforms, shiny white helmets, and hooded eyes. I couldn't imagine them viewing cow-shooting with leniency. I worried about things like this constantly—stepping over the bounds, getting into trouble.

As it turned out, a few weeks after this episode, Dian was summoned to the head of police in the nearest town, Ruhengeri, to account for her actions. She made the long trek down the mountain, the jolting car drive into town and, smoking a cigarette and wearing the blackest of sunglasses, met with the chief. She ended up telling him, "If you don't do your job, then I have to do it for you." I was constantly awed by Dian's boldness and the things she managed to get done.

We didn't have many problems with cattle after that, but they weren't Dian's main concern anyway. The real villains, the Princes of Darkness, were the poachers. Many of these were Twa, a minority tribe in the country. They were small and wiry like the Pygmies from Zaire, and came from a long tradition of forest hunters; only now that the habitat was a Park, they were poachers. They lived on the edge of the Park, cultivating small fields, but still made frequent forays into the forest, sometimes spending several days at a time in the cold, wet mountains. They hunted primarily for meat, eating a small part of the catch themselves and selling the rest.

A few of the more sophisticated poachers had guns, but most of them used spears, bows and arrows, and snares: the spring-traps made from bamboo poles and wire nooses. These caught hyrax, duiker, and bushbuck. Larger quarry, like buffalo, were speared after being run down by dogs who had jangling bells tied around their necks. Dian told me that the poachers smoked *bangi* (marijuana) before a buffalo hunt to give them courage. They were very tough. Because they were so at home in the forest, Dian had a certain respect for them, especially the old guys. But this did not diminish her hate.

Although gorillas were not eaten in this region, they were occasionally killed, especially the silverbacks. Their skulls were sold to tourists, their fingers and genitals were taken for black magic, and infants were captured to sell. But the most common danger to the animals were snares. A gorilla would step into a noose, pull free from the pole, and then walk around with the wire getting tighter and tighter, cutting into the flesh until the limb dropped off or the animal died of gangrene, whichever came first. Some gorillas in the region were missing feet or hands. I could see why Dian hated the poachers. As yet, none of the study animals that she knew personally had been killed, but she figured it was only a matter of time.

This doesn't mean that the threat of poachers darkened each day in the Virunga Volcanos. In fact most of the time it was easy to forget about the dangers. Weeks could go by with no sign of a poacher, not even the print of a barefoot in the mud. Researchers became preoccupied with mundane things like how to keep notes dry in the rain, or with questions about the gorillas and their social intrigues: who was pregnant; who had fathered this new infant; why had this female suddenly run off with a strange male? It was easy to get sucked into their soap operas. At night I dreamt about having conversations with silverbacks.

Each day began with finding a study group, which meant going to where the gorillas had last been seen and then following their trail. The lush secondary growth is so thick on the ground that a group of gorillas leaves behind a swath of flattened foliage that's usually hard to miss. Nevertheless, I was poor at tracking. If the trails of two groups crossed, then I'd follow the wrong group. Sometimes I even followed the wrong species and ended up with buffalo instead of gorillas. If trails from different days were mixed together, I could spend hours and hours staring intently at the ground while walking in large circles. Alternatively, if I was accompanied by one of the Rwandan trackers who worked at camp, it seldom took me longer than an hour to reach the gorillas. Not surprisingly, these men were my heroes. They could follow anything. They once tracked a researcher who was lost by following his faint footprints through a trampled mess of buffalo and gorilla trails, in the pouring rain, at night. They found him at two in the morning.

Once you reached the gorillas, then you stayed with them for a large part of the day, noting the activities and interactions of the animals. You followed them wherever they went, up to the crater lake at 12,000 feet, in and out of steep ravines, through narrow tunnels in bamboo thickets. This called for a wide range of locomotory techniques: scrambling, crawling on all-fours, slithering on the belly—whatever got you there. The gorillas' routine became your own.

Their life was peaceful most of the time because there wasn't much to fight over. The animals had abundant food, living as they did in a gigantic salad. They had no dangerous animal predators. They rarely seemed hurried. Sometimes the whole group would fall sound asleep during a siesta in the middle of the day and I would sit with my pen poised, waiting for some behavior, and maybe nod off for a moment or two myself—one human and thirteen gorillas asleep in the sun in the middle of the forest.

It was easy to think that these animals led a charmed, care-free existence made up of eating, sleeping, and playing. Even the slow dignified adults seemed to have a certain *joie de vivre.* Sometimes during a sedate progression down a slope, they would suddenly somersault instead of walk—like chil-dren at a picnic. It was easy to be lulled into seeing the forest as a lush paradise. But something always happened to snap you back. You would come upon a struggling antelope caught in a wire snare, its leg snapped and twisted like a match stick; or you'd hear the baying of poachers' dogs and the shouts of their masters from a distant hillside. The gorillas could hear it too; they'd freeze for a moment and then panic, fleeing silently in single file behind the silverback. Suddenly there was evil in the forest.

Everyone who worked at camp was expected to put time into anti-poaching work. If you were in the forest following a gorilla trail and you came upon a poacher or their paraphernalia, then you moved out of research mode and into police mode.

Most activity against poachers did not involve guns. When I came to a line of their wire snares, I'd spring them, and take the wire. When I came to one of their small camps, I'd confiscate their spears and *pangas* (machetes) and burn their ragged clothes; I'd tear down their rough huts made from branches; I'd uproot the little plots of marijuana they had planted. In my head I would hear Dian's urgent voice, "Think of the gorillas. Think of Uncle Burt trying desperately to defend his family from spears and dogs." All the same, my heart was never in these activities. Here was some poor guy who had hunted in the forest all his life, like his father and grandfather, and I arrive fresh from Los Angeles, California, and start trashing his campsite.

Besides, I was so incompetent at aggression, especially active pursuit. This invariably ended in ignominy and humiliation. I'd see a poacher in the forest, wave my panga menacingly the way Dian had told me, and begin the charge with a furious yell. Then my foot would catch in a vine, or slip in the mud, or trip on a log and I'd fall flat on my face. I'd imagine the poacher going back to his buddies around their campfire and having a good laugh about it over bushbuck chops and a couple of joints.

Then there was the time I charged at a poacher who was crouched down in some sinister activity I couldn't make out. I got closer and closer, but he didn't run away. *Uh oh.* I had been nervous about this. I had often wondered what I would do if they called my bluff. ("They always run away, Kelly. These people are scared to death of us.") The guy just stayed there. At about twenty yards, I saw his surprised eyes staring at me. Then I realized that I had caught him in an intimate moment, squatting alone in the forest with his pants around his ankles. I was mortified. I might have even yelled out an apology before slinking away. I never told Dian about these incidents. I just prayed that she would not ask me to escalate. I felt I had reached the limits of my militancy.

My direct involvement with Dian's guns first came when she asked me to help her smuggle another Beretta into Rwanda. Dian had a license for her Smith & Wesson which was kept as protection against the feral dogs that roamed wild in the forest and often carried rabies. She had already been bitten twice. The other pistol, the Beretta, was illegal. Now she wanted a third, a later model or something. Just because Dian was bold didn't mean she wasn't paranoid. It was clear that she felt more and more under siege; it was time to stockpile.

I had a friend in Nairobi who took out photographic safaris, but who had previously been a hunter. Because of this, Dian figured that he had access to all types of guns. She came to my cabin one evening with, she said beaming, a "brilliant" idea. Would I write to my friend and ask him to buy a Beretta and send it to her in a cake baked by his wife?

What? Was she kidding? Was this a test?

"Oh, don't look so worried, Kelly. Honestly, it'll be eeeeeasy."

She was kidding. God please make this be a joke.

"His wife could put the cake in one of those big round tins. If they sent it airmail, it would be here in no time. I don't know why I haven't thought of this before."

Oh it's a beautiful idea. Why don't we have them include some ammunition disguised as raisins? Doesn't this only happen in cartoons?

I remained calm.

"Uh, Dian, don't you think this is a bit risky? I mean, what about the customs officials at the Kigali airport? Aren't they going to be little suspicious of such a heavy cake?"

She clearly thought my concern was ridiculous and replied smiling, "Oh noooooo. What about those cakes that English people eat at Christmas? They weigh a ton. Just tell your friend to write 'English Christmas Cake' on the outside of the package. The Rwandans won't suspect a thing. They don't think like us, Kelly. They're different."

"If they're so different, how are they going to know what a fucking Christmas Cake is?" I blurted out and immediately regretted it. Dian's smile disappeared and was replaced with a look of disbelief, followed by deep disappointment. There was a long silence during which she lit a cigarette while her dark eyes bored a hole into me. I began to feel very small. Then she said, "I guess I misjudged you. I thought I could count on you." She shrugged and left, and I spent the following two hours composing the letter to my friend in Nairobi. The next mail-day, after checking the letter out with Dian, I sent it to my friend, praying silently that it would not reach him, which of course it did. We never heard from him, but years later when I saw him in Nairobi, he reminded me of that hilarious joke letter I had once sent him. The one about the pistol in the cake.

I suppose it was a short step from asking me to smuggle gun cakes through the African mail to suggesting that I carry the Beretta into the forest. There were signs of poachers in the vicinity. Nemeye, the tracker, had seen footprints on a buffalo trail; and that morning while out in the forest, I had heard dogs barking and their bells jangling. It was a sound that made the back of my neck crawl.

I was to carry the gun, loaded, in my pocket. If I spotted a poacher or, more probably, a group of poachers, I was to run at them shouting ferociously and firing the Beretta into the air.

"But Dian, I don't know how to use a gun."

"It's eeeeeasy. Here." She showed me how to turn the safety on and off, put eight bullets into the magazine, shove this into the butt after which all eight bullets could be fired one right after the other. She told me to try and hit the empty can of Blue Band Margarine that she had set on a stump about ten yards away, and handed me the gun. It was shiny and black like a bruise; I hated its feel, as cold and heavy as something dead. To me, it was an entirely sinister object and besides, I was an appalling shot. I took aim at the picture on the Blue Band Margarine tin—a smiling African eating a piece of sliced white bread—and squeezed the trigger. I was shocked at the kick of such a small weapon. My hand was knocked back into the air. The man eating his slice of bread was still on the stump, unchanged. God knows where the bullet went. I felt I had no control over the gun: quite the opposite, in fact.

"You see, Dian, I can't shoot to save my soul. I shouldn't be let near this thing," I said, handing the gun back to her. It didn't work.

"You're not supposed to *hit* the poachers; you're only supposed to scare them off. Just fire into the air."

"But I'm such a terrible shot, I might hit them by *mistake.* Besides, don't bullets have to come back down? They might fall onto the poachers' heads because of gravity."

"Well, that's not your problem." Her eyes narrowed. "Look, Kelly, don't worry about the goddamn poachers. Worry about the gorillas. Think of little Titus. Don't you want him to grow into a silverback? Think of Cleo caught in a snare, with the wire. . . ."

"I am, Dian, I am. But for God's sake, this is totally illegal, isn't it? What am I going to say if I run into a guard patrol from Rwanda or Zaire and they find this gun on me?"

"These sons-of-bitches don't go on patrol, Kelly. That's just the problem." She was beginning to look disappointed, but I said nothing. Finally she sighed, still fixing me with her dark eyes. "It's not your fault. It's mine. I should never have had a girl come to do this kind of work."

The gun felt like twenty pounds in my jacket pocket as I set off very early the next morning on my way to Group 4. I thought it would just get heavier and heavier, spoiling my day, but the morning was so clear and beautiful. On the western horizon, jagged Mount Mikeno stood flushed with pink against the pearly sky. A bushbuck doe crossed the trail ahead of me, lifting her exquisite legs deliberately with each step. A sunbird of iridescent green hovered above a lobelia spike.

It was a late, lazy morning for the gorillas. Seven in the morning and they were still on their night nests. Most of them were hunkered down, not quite awake yet—black shapeless mounds surrounded by dewy greenery. One by one they yawned and stretched, looking rumpled from sleep. Gradually the cold quiet air grew noisy, filling up with the hums and grumbles, with the rustling and shifting of a group of gorillas about to start the day.

I always felt like I was being let in on something very intimate and private when I was with the gorillas on their nests. I thought, Dian's right. Maybe you do just have to make a choice, put your head down, and just go. Black and white. My hesitancy and inaction suddenly seemed so pathetic.

I spent six hours with Group 4, during which it was hot and sunny for a time. It got dark and freezing clouds suddenly poured into the ravines with a rushing sound. Then there was a tremendous hail storm and the gorillas sat hunched over with their arms folded across their chests while the hail piled up in little white mounds in the crooks of their arms. Then it poured with rain for an hour. Then it went back to warm and sunny and the animals stretched out like sunbathers, steam rising from their bodies. Individual raindrops clung to their fur and sparkled like tiny jewels. A mother suckled her infant, Uncle Bert mated with Flossie, and there were a couple of minor squabbles. I left the group moving through the dripping shrubs and vines, eating hungrily.

I had completely forgotten about the gun. In fact, I didn't think of it until I was only about twenty minutes away from camp. I was ambling along the meadow when suddenly, something caught my eye, about sixty yards away in the woodland; a small wispy cloud, hovering just above the ground. At first I thought it was mist, but no such luck. It was smoke. I noticed a shadowy human shape moving among the trees, and then another. They were poachers, building a campfire. *How dare they relax so close to camp!!* Then, I remembered the gun.

I reached into my pocket, took out the Beretta and switched off the safety latch. The little red dot jumped out at me like a warning spot on a bug. I took a deep breath. *This is it. Here we go.* Lifting the gun into the air with my right hand and giving my most blood-curdling yell, I began the charge.

The morning rain had turned the meadow into a bog. As I ran, up to my ankles in mud and water, I made great sploshing noises. It was like running across a giant, sodden sponge. I stumbled, but kept going. With the gun barrel pointed up at the sky, I was just about to squeeze the trigger when I noticed, in the murky shadows of the wood, that there were rather a lot of figures. Some were sitting, others standing, others now looking towards me. This was not the typical "handful" that usually made up a party of poachers.

In the next two seconds or so, many important things occurred to me. I noticed that some of the figures were wearing boots, *but poachers don't have shoes;* and those long things leaning up against the trees were *rifles, not spears.* I realized that these guys were all dressed alike, *and they were wearing berets.* By now my battle cry had strangled itself and my right hand with its gun was back in my pocket. My splashing gallops had slowed to long, awkward strides, but on I plunged across the boggy meadow smiling as hard as I could. Having dropped my panga, I was now waving with my left hand.

Unconsciously, but hardly missing a beat, I had transformed my assault into an inanely friendly welcome at the moment it became apparent that these were not poachers, but the Commandos, Rwanda's ace military unit. Their relentless training included periodic stints of running up and down the Virunga Volcanos, and here they were, fifty of them, all standing and watching my approach. Squelch, splash, squelch, splash. I yelled, "*Bonjour,*" with a faint tremor of hysteria, feeling sure that all eyes were on my right pocket which was swinging heavily against my hip in a gun-like fashion.

The commander stepped forward to meet me. He was a squat, thick-necked Belgian with a blond crewcut, and a face as red as a brick. I was panting and shaking and saying "*Bonjour, comment-allez vous, beaucoups de soldats, n'est-ce pas?*" and other random French phrases that came to mind. He was suspicious. Who was I and what was I doing here? Did I have permission to be in the Park? Ah, yes, Dian Fossey's research camp, he had heard of it. Should I be out here alone? It was dangerous. There were wild animals around. Before I knew it, we had said "*au revoir*" and I was heading for camp. It was over.

As I followed the trail along the creek, I thought of how close I had come to firing on fifty Rwandan Commandos. No doubt they would have returned fire first and then asked of my bullet-riddled corpse, "*Mais, qui est cette fille?*" I began to feel fragile and shaky, like I had the flu. By the time I reached camp, I felt quite ill, but resolute. I walked into Dian's cabin, handed her the Beretta and said, "Dian, don't ever ask me to carry this thing again." She didn't argue. In fact, she was very friendly. She gave me a beer and asked me about the gorillas. We chuckled over my assault on the Commandos. I think she knew I would never be a Joan of Arc. Actually, I think she was a little grateful.

The Great Parrot Hunt

by James Serpell

THE EVENTS SURROUNDING the great parrot hunt occurred in 1975 during one year of fieldwork in Australia and Indonesia. The main object of this research was to investigate the natural behavior and ecology of a genus of small, nectar-feeding parrots or lorikeets. This genus, *Trichoglossus*, is intriguing for several reasons. First, it is of recent evolutionary origin. Second, it consists of a large number of isolated species and subspecies, each of which has evolved its own unique, and often highly ornamented, plumage coloration. Third, the group as a whole performs a remarkably histrionic repertoire of visual displays. My ultimate purpose was to study the variation in these visual displays much as one might study the evolution of different language groups in humans. But, in order to do this, I needed representative groups of wild-caught lorikeets to observe in detail in captivity. For this reason, I devoted a considerable portion of my fieldwork to the business of locating, capturing, housing, feeding and, eventually, exporting lorikeets from a number of outlandish locations. After ten months of uninterrupted travel, I arrived eventually in Ruteng, a small town on the mountainous, Indonesian island of Flores. And here begins my tale.

In addition to being immensely tall, the Dutch missionary, Father Verheijen, was an authority on Wallace's Hanging Parrot. This was an interesting achievement, since Wallace's Hanging Parrot was known only from a single museum specimen and had never been observed in the wild, either by him or, as far as I could tell, anyone else. Mind you, Hanging Parrots are not always easy to see, owing to their habit of impersonating leaves while dangling upside down in the tops of tall trees. (Some weeks before, in Sulawesi, I had witnessed the uncanny sight of a tree full of Hanging Parrots suddenly and spontaneously defoliating itself when I disturbed a large roost.) Despite failing to confirm the existence of this mysterious bird, Father Verheijen had nevertheless assured me in his letter that the parrot I was seeking—Weber's Lorikeet—was numerous on the island of Flores, and especially so near Ruteng.

"You must ask in the villages," he said when I arrived. "They eat them you know. They eat everything here."

Sure enough, tentative inquiries at the nearest village provoked an enthusiastic response.

"*Burung ini banyak sekali disini*" [This bird is very common here], explained the headman. "Every night he flies up the mountain to sleep. In the morning he flies down again. These men," he said, with a gesture that encompassed the entire village, "will show you this bird." In the gathering dusk, we set off down the track accompanied, it seemed, by every able-bodied man and boy in the community. Eventually we came to a large stream which meandered down the mountain side, flanked by trees, long grass, and tussocks of bamboo. At a signal from the headman, everybody dispersed into the undergrowth amidst a great deal of laughter and shouting. Suppressing a growing feeling of skepticism, I crouched down beside him and together we squatted quietly in a thicket.

Then, after twenty or so mosquito-infested minutes, the silence of the evening was interrupted by a series of thin, metallic shrieks from further down the valley. The headman seized my forearm in a vise-like grip and whispered urgently, "*Dia datang!*" [They are coming!]

The shrieks drew rapidly closer, and seconds later a sizable flock of Weber's Lorikeets whirled past us on their way uphill using the course of the stream as their flight path. No sooner had they gone than another flock could be heard in the distance approaching by the same route. And so it went on, a sort of psittacine rush hour, until darkness finally descended. Judging from the excited smiles and general chatter which followed, I was not the only one to be impressed by the accuracy of the headman's forecast. Even he seemed to be a trifle overwhelmed by it all.

"Tomorrow we will catch this bird," he announced confidently. "This bird is very good . . . delicious!"

On the journey back to the village the men attempted to explain their methods for catching lorikeets. The procedure, which appeared to involve swatting the birds with bamboo poles, sounded too implausible and lethal for my purposes, so I revealed that I possessed a remarkable net which would certainly capture the parrots alive and unscathed. This information was greeted with discernible suspicion.

"How big is this net?" they inquired dubiously. "This bird is very clever. He will see the net and fly round it." My Indonesian was not up to the task of describing a Terylene mist net, and in the end I was obliged to leave them wondering about the mysterious, supernatural properties of a net that was simultaneously huge and invisible, and yet small enough to fit in my pocket.

Dinner with the missionaries that night was full of surprises. I was prepared for an evening of austerity—prayers of thanksgiving followed by a meager repast, water or perhaps fruit juice to drink, and early to bed. Instead, the meal was long, sumptuous, raucous and convivial, and punctuated by risqué jokes. After the food and wine, the Abbot produced a stoneware bottle of aged Dutch gin which smelled of old socks and tasted like fire. This soon disappeared and I took the opportunity to start making my excuses about feeling tired and needing to get an early start.

"But you haven't sampled the local brew yet," slurred the Abbot, "the arak here is excellent!" An unmarked bottle of poisonous liquor appeared and disappeared almost as quickly as the gin. By now the conversation had become dangerously incoherent, and I began to wonder if I would be able to stand up without falling over. With a tremendous effort I lurched to my feet, muttered something about falling asleep, and staggered out into the compound. It was pitch dark and a light rain was falling as I groped my way back towards the little room I had been allocated. But I forgot about the monsoon drain.

Suddenly the ground disappeared beneath my feet and I pitched headlong into a deep concrete trench containing about three inches of freezing cold water. A blinding pain exploded in my right foot. With difficulty I clambered out of the drain, cursing and groaning, and crawled through the mud until I was able to regain my feet—or at least one of them—with the aid of a small tree. The excruciating pain was now more localized around my big toe which, in my fevered imagination, I envisaged pulsating hotly like a cartoon appendage.

"You've broken your bloody toe, you stupid moron!" I thought to myself, "now you've really done it!" Stricken with pain and self-pity, I shambled and hopped the remaining distance to my spartan sleeping quarters. Gingerly I removed my sandal and scrutinized the offending digit. It looked just the same as the other one but it hurt like hell. Exhausted and only semi-sober, I undressed and collapsed on the cot. My body felt unbelievably heavy, and while the throbbing agony in my toe was undiminished, it began, in a curious way, to recede a considerable distance from the rest of me. It was as if my legs and torso had become impossibly elongated, or my foot had somehow detached itself and moved politely to another part of the room.

"Must be the arak," I thought, as I finally dozed off.

The rest of the night passed in a sort of deranged nightmare. At intervals the toe would haul me back to consciousness and I would be dimly aware of feeling cold, bruised, and sick. Then I would lapse back into feverish dreams of leering missionaries and shrieking lorikeets. At one point I awoke to a subtle change in the quality of pain emanating from my toe, as if it was being gently scraped with a scalpel. For a while I lay there stoically accepting this new discomfort until I realized that it was coming from the wrong foot. I sat up and groped for the light switch and discovered, to my utter horror, an immense cockroach quietly devouring the horny skin of the other big toe. With a grunt of disgust, I kicked out violently hurling the repellent creature backwards through the air and onto my face where it immediately sprang to life and scuttled under the bed-clothes—"Uuurrgh!"

Despite my various handicaps, I leapt from the bed like a seasoned gymnast, simultaneously flinging my assailant onto the floor. It stood there for a moment, poised and alert, testing the air with its long, whip-like antennae. Then, with remarkable prescience, it raced for the crack under the door. All at once I was consumed with hysterical rage.

"Bastard!" I screamed, as I watched this violator of my person rapidly escaping to undeserved freedom. With a kind of Zen determination I grabbed one of my sandals and hurled it with extraordinary speed and accuracy at the fleeing invertebrate. Their paths converged at the base of the doorway. The heel of the sandal descended in a final, graceful arc on the roach with a loud and profoundly satisfying crunch.

Perhaps fortunately, many details of the following day have been permanently erased from my memory. A cursory medical examination by one of the missionaries revealed that the injured toe was bruised but unbroken. The nail was slowly turning black and it was painful to walk on, but I was thankful for small mercies. I limped about town during the day arranging the construction of a neat wooden parrot cage and then, at the appointed time in the evening, met the village men beside the stream. Preparations for the great parrot hunt were already well underway. Some of the hunters were in the process of hacking a broad channel through the grass and bamboo which connected two adjacent bends of the stream. The lorikeets, they assured me, would use this shortcut rather than follow the longer curve of the stream bed. They were all agog to see the mysterious net and were far from disappointed by the reality. As we erected it across the upper end of the channel, they gently caressed and admired the fine black filaments, shook their heads with astonishment, and muttered to each other in hushed, respectful tones. The headman stood back and examined the contraption with a speculative eye.

"What do you call this net?" he asked. Anticipating such a question, I had taken the precaution of looking up the Indonesian word for "mist" in a dictionary, but my rendition of "mist net" only elicited a series of puzzled looks. "Where did you get this net?" continued the headman, impatiently.

"In England," I replied. *"Ah! Jaring Ingris!"* [English net!]

Despite the existence of the "English net," several hunters had manufactured a number of low-tech alternatives which resembled overlong lacrosse sticks—bamboo poles about twelve-feet long, the ends of which had been split down the middle, splayed apart, and strung like tennis rackets with further strips of bamboo. As one of the hunters graphically demonstrated, the idea was to lurk in the grass at the side of the channel and then raise the pole swiftly and whack the birds out of the air as they hurtled past. I pointed out that even if this bizarre technique worked, it would undoubtedly kill or injure the parrots. But at this they only smiled evasively and assured me that this particular parrot was too strong to be killed in such a way. I began to suspect a serious conflict of interest but it was too late to abandon the scheme now. Dusk was fast approaching, so we took our places on either side of the artificial flight path and waited.

The advance party of lorikeets consisted of only four birds. As promised, they diverged obligingly from their normal route and came whizzing up the middle of the channel straight towards the net. Regrettably, several hunters could not contain themselves and swung enthusiastically at the birds with their bamboo poles before they got there. As a result, only two tumbled into the net while the others flew off at a tangent screaming loudly. I quickly extracted the two captives and slipped them into cloth bags. Wishing to avoid further escapes, I instructed everybody to stand behind the net and, if necessary, to use their sticks to knock down any birds that flew over the top. I was not fully prepared for the events which followed.

Minutes later a full-sized squadron of some thirty or forty bright-green parrots came swooping up the channel straight into the net. Overcome, I suppose, with the excitement of the moment, three or four hunters immediately lashed out with their poles and these instantly became horribly entangled in the mesh along with the struggling, screaming birds. From here on, things became decidedly chaotic. Before I could stop them, everybody rushed at the net and began grabbing mixed handfuls of netting and lorikeets. The headman, whose wrists and forearms were festooned with ornate bracelets, also got caught up in the netting and began shouting and thrashing his arms, thus adding to the general confusion. Now and then, a lucky bird would escape only to be struck down on either side by wildly flailing bamboos. It was like being trapped in the middle of a demented badminton tournament. I managed to extricate half a dozen birds without too much difficulty and bundled them into bags. But the rest were so horribly enmeshed that my heart sank. There is an art to removing birds from mist nets at the best of times. It requires patience, calmness, and manual dexterity. Usually, one is only dealing with one or two birds at a time, instead of twenty or thirty, and the random assistance of large numbers of inexperienced helpers is generally considered undesirable. On top of this, parrots do not react to capture like other birds. Most birds remain relatively passive once entangled, but lorikeets tend to wriggle and squirm and fight with the net until they either escape or become completely cocooned in an intricate rat's nest of black Terylene. And being parrots, they also bite. There seemed to be no alternative. I got out my penknife and began cutting my precious mist net to pieces. By the time I had finished, both the net and my fingers were in shreds and, while some birds were safely liberated, a depressing array of feathery, green corpses lay on the ground.

As far as the villagers were concerned, I was a hero. As we walked home through the darkness, they clapped me on the shoulder and recounted the evening's events to each other in glowing terms. I later calculated that the headman offered me the equivalent of three months' wages for the remains of my "English net" but, imagining the ornithological devastation it might cause even in its present condition, I refused to sell. He accepted this with equanimity, and invited me and several of his senior henchmen into his house: a large wooden bungalow on stilts overhanging a ravine. Inside we arranged ourselves on raffia mats on the floor and were served sweet, milky tea in glasses by his small and beautiful wife. Meanwhile the rest of the villagers assembled outside on the verandah where they peered in eagerly through the windows, their eyes and faces shining warmly in the dim glow of kerosene lamps. Some of the bolder lads demanded to hear the story of the "white man and his English net," and our host was more than willing to oblige. His gold teeth flashing in the lamplight, he began to re-enact scenes from the great parrot hunt, illustrating each episode with animated gestures and histrionic sound effects. Every so often, other hunters would interject missing details and these would set him off on another train of vigorous reminiscences. The overall effect was mesmerizing. When he came to the part about getting his own bangles caught in the mesh, he brought the house down. The entire village shrieked with laughter, and he himself was rendered so helpless with amusement that I began to fear for his health. His wife eventually brought him around with more tea, but it was some time before he could continue his account without breaking off into further tearful paroxysms of mirth.

At the end of the story a sort of thoughtful silence descended on the company, and I began to think about making my departure. I was just about to rise to my feet when there was an extraordinary noise like a tiny aircraft engine, and a large black object sailed through the window, crashed into the lamp, and fell with a loud thump on the floor beside me. It was a huge rhinoceros beetle with immensely long, forward curling horns. To my surprise, its arrival was greeted with loud oaths and general consternation. But this was nothing compared with the howls of anxiety when I reached over, picked the creature up, and placed it on my hand for closer examination. The headman flapped his arms at me anxiously.

"Don't touch it! This animal is very bad. It will bite you!" Everyone was so alarmed that I started wondering whether my rudimentary knowledge of rhinoceros beetles was seriously deficient. I tried to remove the beetle, lifting it by its horns, only to discover that the harder I pulled the deeper its claws imbedded themselves in my already lacerated fingers. For an insect it was unbelievably strong and immovable, and its claws were like needles. Feeling somewhat embarrassed and not knowing what else to do, I went out on the verandah and tried shaking it off. This caused me considerable discomfort and failed to dislodge the beetle. Tiny beads of blood appeared where its claws had punctured the skin. I felt completely ridiculous and was on the verge of accepting it as a sort of bizarre accessory when, at last, it slowly unfurled its wings, detached itself from my hand, and roared off into the night.

I returned to the room, smiling somewhat lamely, and attempted to make light of the whole incident. But the general atmosphere had definitely turned sour. The rhinoceros beetle seemed to have cast an ominous pall over the entire proceedings. In an effort to restore the earlier mood of hilarity, I decided to recount my previous night's debacle with the cockroach. This also proved to be a mistake. Whether it was the taboo subject of insects, my limited Indonesian vocabulary, or the manner of storytelling, I will never know. But whatever it was, I failed to raise so much as a smile. On the contrary, each blow-by-blow description—the fall in the drain, the bruised toe, the man-eating roach, the leap from the bed—evoked looks of sincere horror and various gestures of condolence. Only the roach's final demise seemed to provoke any satisfaction at all, and even then it only produced a grave nodding of heads, as if I had inadvertently revealed some profound moral truth.

"This animal is very bad," concluded the headman sagely.

Fortunately, I was saved from further social gaffes by the arrival of food—a large steaming basket of boiled rice and a delicious spicy-smelling stew. "*Makan!*" [Eat!] said the headman, offering me a bowl and spoon. Gratefully, I piled some rice on my plate and delved into the rich concoction. As always, the sauce was heavily spiced with chili and there were pieces of rather fibrous meat floating around in the mixture.

"*Enak!*" [Delicious!] I said, pausing to remove a mouthful of small bones which I examined briefly before placing on the side of my plate. They resembled tiny chicken bones. Over in the corner, I glimpsed the captive lorikeets eyeing me accusingly through the slats of their cage. I began to experience a sudden loss of appetite. "Umm . . . What kind of meat is this?" I asked the headman hesitantly.

"This?" he replied, somewhat startled by the stupidity of the question. "This is lorikeet."

~

The walk back to town later that night gave me time to reflect on a day of dubious achievements. In the space of twenty-four hours I had been drunk under the table by missionaries, terrorized by a cockroach, and acutely embarrassed by a beetle. I had fallen in a monsoon drain, injured my foot, decimated the local avifauna, ruined an otherwise amusing dinner party, and unwittingly devoured at least one of the animals I was supposed to be studying. As far as I could judge, I had also created an insatiable demand for "English nets" which, if ever satisfied, would probably reduce the island of Flores to an ornithological desert in a very short space of time. Aside from the box of angry lorikeets under my arm, my only consolation was a curious and unorthodox insight into the primitive origins of racket sports.

The Stubborn Snake

by Monica Udvardy and Thomas Hakansson

THE STORY WE are about to relate concerns a class of animals to which most humans display a certain irrational fear. This is a story about snakes—or rather, a particular snake. . . .

Out of fortuitous happenstance, both of us share many interests, one being respective careers as Cultural Anthropologists. Thomas, however, has one passion which Monica definitely does not share. It is a hobby that he particularly likes to indulge in when he is in the field—a habit which has several times put our relationship under considerable strain. That interest is herpetology: the biology of reptiles and amphibians.

Having seemingly been born with this interest, Thomas has difficulty understanding that many people—most, in fact, do not appreciate the charm of snakes and other reptiles. Having reluctantly acknowledged that Monica belongs to that peculiar majority, Thomas tried to woo Monica to the cause. Some of the more innocuous representatives of the class, such as tree frogs and lizards, encountered little opposition, and with time, Monica has even been overheard to remark that some of these are "cute." Snakes, however, are another matter. Thomas has tried to capture her interest by relating the amazing and remarkable traits unique only to this kind of reptile. And to her repeated inquiries about the dangers of handling snakes, he normally scoffs, boasts about how harmless they are, and ends by emphatically stating that they (almost) never bite. . . .

Our story takes us to Kenya, where we conducted fieldwork for our Ph.D. dissertations in phases between 1982 and 1986. A persistent problem faced during that period, and which indeed is common to most contemporary, dual-career, and internationally-oriented couples, was how to combine our personal relationship with the demands of our profession. Thomas was carrying out research in western Kenya, among the Gusii people and Monica was conducting her fieldwork among the Giriama, a coastal hinterland people located about 500 miles away from the Gusii, on poorly surfaced roads. In order to minimize the distance between us, Monica purchased a used Toyota Land Cruiser, which she affectionately christened the "White Elephant," both for its powerful motor, and for its copious consumption of petrol. Fortunately, few long-distance commutes in this vehicle were necessary since Thomas found enough herpetological interest in the lush and gently rolling, humid ridges of the coastal inland among Monica's Giriama villages to justify long visits to the coast. We were able to synchronize our respective work so that after the initial three months, Thomas joined Monica near the coast, after finishing his own field research in Gusiiland.

Thus, in 1985 we were staying together in the small Kenyan trading center of Kaloleni, situated amidst coconut palms, marshes, bush, and small patches of forest, some eighteen miles northwest of the port of Mombasa. Monica had already converted a small, local tea house into comfortable, if crowded, living quarters. Our mud-walled and palm-thatched house was also home to an entomologist's paradise of insects, both scuttling and aviatory, as well as chattering geckos, scurrying rats, and giant millipedes, the latter locally dubbed "Mombasa trains" for their enormous length of six to twelve inches.

Monica was conducting fieldwork to learn about the lives of older Giriama women, specifically post-menopausal women. In this patrilineal society, where men, and especially older, well-respected men, are awarded power and authority simply by virtue of *being* men, Monica wondered, "How do older women gain power and influence in their families and communities?" The answer to this question involved searching out the relatively few elderly women amongst the large, extended and very child-centered families that characterize Giriama domestic life.

The Giriama were remarkably gracious and hospitable toward Monica, and soon older women were clamoring for visits by this peculiar white person (or *mzungu,* as she was initially referred to in Kiswahili). At first, the lightness of her skin, the smoothness of her hair, and her utter inability to carry all manner of objects on her head, as all Giriama women do effortlessly from the age of five, provided robust live entertainment for large circles of family and friends of all ages.

Before Thomas joined Monica, she was not lonely because, despite her apparent eccentricities, the Giriama women befriended her. Initial amusement gave way to more serious thoughts, the most important being: "Why is this peculiar woman alone, without a husband?" "Why isn't she married?" And if (as Monica claimed) she was married, then *"Where was her husband?"*

The concern of these kindly women gave way to generous offers that she be married to their own husbands. As is common throughout Sub-Saharan Africa, the Giriama practice polygyny. Co-wives among the Giriama generally get along well, the younger ones relieving the workloads of their elder co-wives by taking on the heaviest work tasks, such as carrying water and fuelwood from great distances (on their heads), pounding maize in mortar and pestle, and grinding it into flour. These offers were poignant in light of Monica's soft and underworked arm and thigh muscles and her obvious inability to carry almost anything on her head. Eventually Monica's nickname was changed from one that marked her as different, to one of affectionate, fictive kinship. Until Thomas arrived on the scene, women tenderly referred to her as *mkaza*, or co-wife.

After the initial three months, in which Monica spent her days (and sometimes nights) roaming the hills in her quest for data, Thomas, the long-awaited husband, finally arrived. He immediately and enthusiastically took up his favorite hobby of collecting snakes, frogs, lizards, and other sundry varieties of reptiles between sedentary bouts of working up his Gusii ethnographic material. As scientific cover for this immensely enjoyable hobby, he contacted the administrative body of the National Museums of Kenya, and gained their immediate approval to compile photographs and build up an assortment of coastal reptile specimens for their collections.

His first task was to build up a "snake information network." Word spread that this peculiar European would reward children of all sizes with candy and soda in return for on-the-spot information on the whereabouts of the more unusual local reptiles, especially snakes. As the children caught on, sightings flooded in. We would awaken to the tapping of small hands on our door, and young children impatiently waiting to drag Thomas to nearby huts and houses. Soon he was overwhelmed with requests to rid neighborhood residences and local farming plots of much feared reptiles, and spent his days (and sometimes nights) extracting loitering green mambas from house beams, spitting cobras from pit latrines, and all manner of poisonous and nonpoisonous snakes from farming fields.

Local Giriama marveled at this queer European male, often spotted strolling systematically up and down fields and bushy areas, head bent towards the ground as though deep in thought, clad only in shorts, knee-high rubber boots, special, heavy leather gloves, and poking the bush with a strange aluminum stick: the snake-handling tool of every serious herpetologist. Made of highest grade airplane aluminum for lightness of weight, it is hooked on one end, in order to pin the head of a venomous snake firmly to the ground. Then, with head securely out of arm's reach, the careful herpetologist quickly grasps the long, writhing body and, in one fell swoop, thrusts the snake into a thick canvas bag. These bags have heavy-duty, drawstring openings that can be securely tied with a couple of twists before the live animal has time to react.

Thomas realized early on that he needed to have a cage to store the snakes until they could be properly recorded, photographed, and freed in remote areas. The snakes more than earned their keep. Since our easily penetrable, palm-thatched roof did not leave our house safe from burglars, we needed to devise security measures. The solution was obvious: a "watch" snake. Our cage of live snakes was prominently displayed outside the house, and the security system worked perfectly. Surely the Giriama wondered what kind of husband this was, who routinely handled this box of animals which are among the most feared. While too politic to broach the subject of her choice in spouses, Monica's neighbors and friends did coin a nickname for Thomas: *mganga wa nyoka,* roughly translated as "snake doctor" or "snake magician."

In addition to serving double duty as scientific collector and local pest controller, another dimension of Thomas' self-appointed task was to search for herpetofauna during concentrated evening forays in the local swamps and remnant rain forests, outfitted with a powerful miner's headlamp. Not altogether convinced of the wisdom or safety of these nightly sojourns, Monica nevertheless assented, and lent him her precious White Elephant. She also lent him one of her assistants, Shaban, who was required to carry the cumbersome equipment and explain to late-night pedestrians encountered on dark paths that the strangely-attired European was only a harmless lunatic and not a wizard about to commit sorcery in the dead of night.

One evening, a series of remarkable circumstances were set in motion that put a dent in Thomas' smoothly-running snake-collecting operation and, more seriously, into the fragile public relations campaign for snakes that he had so carefully nurtured with Monica.

That night, Thomas and Shaban parked the car close to the forest and approached a cultivated field near its perimeter. The field was densely planted with bananas and cassava. On the ground, the farmers left piles of fresh banana leaves, ideal hiding places for frogs or small snakes. Thomas began turning over the leaves one by one with his hands.

He suddenly glimpsed a small black snake under an overturned leaf. Instinctively, he grabbed it with his left hand and, as it whipped its head around in a uniquely oblique way, he realized as it sank its left fang into his flesh that he had made a classic mistake! He felt a sharp pain as if being stung by a wasp, and realized that this was not a harmless Purple Glossed Snake

(Amblyodipsas), but a poisonous Southern Stiletto Snake *(Atractaspis bibroni)!* The two are almost identical, and anyone working with snakes in Africa is repeatedly warned in the literature that these two are easily confused. The Stiletto Snake is nocturnal and spends much time searching for rodents in their burrows. It has very long fangs sticking out Dracula-like on the sides of the mouth. By whipping its head from side to side, it can effectively puncture the skin of its prey or, alas, the amateur herpetologist. Like most vipers and rattlesnakes, the venom, largely hematoxic, affects the tissue by causing swelling, necrosis, and internal bleeding. It is not deadly poisonous, but a bite in the hand gives rise to a nasty and painful swelling which subsides after a couple of days, leaving the limb discolored and aching.

Jolted by the bite, Thomas shook his hand so that the snake could not strike again, but clung to it all the same, not wanting to lose this valuable specimen. As he deposited the snake into the cloth bag which Shaban was holding, his hand began to swell. Tying the drawstring, his fingers moved slowly, having swollen to the size and color of medium-sized Polish sausages. For this reason, he canceled the remainder of the evening's plans and went directly to the local clinic, where he hoped to get some medicine to counteract a possible allergic reaction. Because the specimen was only fourteen-inches long, injecting an anti-venom was unnecessary. The snake in its bag was placed in the car and they quickly drove to the clinic. After collecting some allergy tablets, they headed home.

Thomas entered the house with his left hand behind his back, and after a casual recounting of the evening's catch, he slowly withdrew his hand. Monica was by no means pleased, but his sorrowful condition elicited a good deal of sympathy and worry and, after some discussion, we decided to go to bed.

As Monica headed for the bedroom, Thomas went to fetch the snake from the car and deposit it safely into the cage.

After opening the car door, he grabbed the cloth bag which seemed suspiciously light. Shocked, he found it was empty! The snake had somehow managed to escape! In the car! Escaped in Monica's car—the White Elephant! Thomas must not have been able to tie the bag's drawstring properly with his swollen fingers. Now there was a poisonous snake loose in the car.

Speechless, he collected a flashlight and began to search. But the night was dark, the snake was black, and so was the vehicle's upholstery. By now, Monica had come out to inquire why he was taking so long, and he was forced to relate the disturbing situation. Grimly, Monica helped in the search, but after a few minutes we gave up and decided to continue hunting after daybreak.

~

Early the next morning, the search continued. That is, Monica and the two field assistants searched, as Thomas sat directing the activities from the comfort of an armchair—his arm too swollen to allow for any strenuous activity. We knew that the snake had not left the car, as the sand underneath revealed no telltale slithering patterns. After a futile, superficial search, we had no alternative but to begin to dismantle the car's interior. Rubber floor mats and other loose items were removed, followed by the car seats themselves. The petrol tank, located under the passenger seat, was also removable. As Shaban lifted it, Monica poked a flashlight underneath, and, just for a moment, glimpsed the shiny eyes of the snake! Finally! But when we hurriedly removed the tank, the snake had disappeared once again. We searched until three in the afternoon without success, and reluctantly gave up.

In the doorframes of the Land Cruiser, we had noticed rows of small holes near the bottom, largely plugged. The holes were probably intended for water drainage, but since some of the stoppers had long since been lost, we concluded that the snake must have crept through one of these holes, and now lay lodged between the inner and outer frames of the door. Presumably, if we plugged these holes, the snake would not be able to escape, but would be left to die. The solution seemed to be to plug the holes.

Having accomplished this, we reassembled the car. Frustrated at not having found the snake, we were left with a persistent feeling that the snake might make an unexpected appearance at some future, inopportune moment.

The following day we decided to take a little time off. We packed the car and left Kaloleni, heading for the beach north of Mombasa, and the prospect of a lazy day lounging under palm trees to the music of gently lapping water. We drove the fifteen miles down the tarmac road from Kaloleni until reaching the Nairobi-Mombasa road, and then continued towards Mombasa. As Monica drove, Thomas kept his swollen, slightly bluish-green hand and arm above his heart by resting the limb on the back of the car seat.

As we reached the perimeter of the city, the road divided into two lanes for a few miles, and we could feel and smell the hot, moist, and salty sea air. This nice bit of two lane road permitted some speed and Monica stepped on the accelerator. Suddenly, amid the heat emanating from the motor, she felt a cool, slimy sensation on her right leg. Before she could comprehend it, the sensation spread to the other leg. *The Snake!* It was crawling over her ankles! Screaming loudly, she jerked her legs away from the peddles and up over either side of the steering wheel. Thomas threw himself under her legs while she desperately tried to keep the car on the road.

"I've got it!" Thomas screamed, as he managed with the help of a wrench to pin the head of snake to the floor between the brake and the clutch pedals. Weaving madly, Monica brought the car to the side of the road, still steering between straddled legs. The car came to a halt amid a crowd of roadside bystanders who watched the suicidal swaying of the car with gruesome fascination.

What this crowd saw next elicited shouts of amazement and perplexity. The driver's side door of the car was thrown open and for a moment nothing happened. Inside, Monica had attempted to throw herself out, but was trapped by her safety belt. A moment later it was unfastened and she cast herself from the car, cursing loudly. Thomas remained in the same crouched position, snake still pinned by the wrench.

"Kill that snake!" she screamed.

"But. . . ."

"Kill it!" she insisted.

"But, I want to keep it alive!"

"Kill that snake!!"

Reluctantly, Thomas pressed down on the wrench and took the life of the creature. The crowd cringed as the headless snake was flung from the White Elephant on its final journey to mother earth.

The awestruck crowd was now treated to a downcast European man who emerged from the car holding a bloody and discolored hand gingerly in the air. His previously bitten finger had ruptured in the struggle and now oozed a messy stream of pus and blood. The surrounding spectators withdrew for a moment in horror, only to surge forward again with murmurs of sympathy. Naturally, they concluded that Thomas had just been bitten, and offered hurried suggestions of help with a tourniquet or guidance to the nearest hospital. In the meantime, Monica, the strangely unsympathetic spouse, was pacing around the White Elephant, inhaling deeply on a cigarette taken from Thomas' pack—the first smoke she had had since her teenage years. Thomas was left quite alone in his attempts to explain to the crowd what had happened.

~

Several whiskies and gin and tonics later, as we lay in our lounge chairs, both of us saw the humor in the spectacle we must have presented. Still, this snake effectively destroyed Thomas' snake-handling reputation, and Monica banned snakes from the White Elephant. Thomas found this prohibition quite unreasonable, and a heated disagreement could have resulted, had we not come to an amicable compromise. Henceforth, Thomas was only allowed to carry snakes in a sealed box tied to the roof rack. Knowing that he would now have to endure a long and painstaking period in order to restore his badly damaged reputation, Thomas pointed out to Monica that this incident really had much redeeming value. It was an unusual adventure that had added excitement to the otherwise humdrum routine of fieldwork, he claimed. And furthermore, it made for one helluvah story!

Section II: Physical Dangers

Bushmaster in the Bidet

by Richard O. Bierregaard, Jr.

RAIN FORESTS ARE in retreat. Human populations in the countries that harbor tropical rain forests are exploding at a frightening pace. Driven by the most basic need of all—a place to live—the homeless and unemployed are moving into primary rain forests. And in areas where massive colonization is not underway, international businesses are stripping vast regions of virgin forest, turning pristine wilderness into particle board and cardboard shipping boxes for our VCRs.

The onslaught of chainsaws, bulldozers, and dams is overwhelming the richest terrestrial ecosystem on the planet. In the early 1970s, Tom Lovejoy, a prominent conservationist, then at the World Wildlife Fund, realized that our ever-expanding population would continue to destroy these precious reservoirs of biological diversity. Logically, he concluded that the rain forests' retreat had to be carefully planned. The goal would be to leave as much rain forest standing as possible and to assure that those areas set aside would maintain their ecological integrity over long periods of time.

Biologists and government bureaucrats charged with protecting rain forests found that data on the relationship between the size of a rain forest reserve and its ability to support the ecosystem it was established to protect were lacking. Concerned about the need for such information, in 1979

Lovejoy launched the Minimum Critical Size of Ecosystems Project, north of Manaus, Brazil, deep in the heart of the Amazonian rain forest. The goal of the research project was to provide critically important data on how best to manage these reserves.

The design of the Minimum Critical Size Project, now called the Biological Dynamics of Forest Fragments Project, is simple. Its execution was not. Project researchers trekked off into areas of the virgin rain forest slated for clear cutting. There they marked off forest tracts of different sizes. By arrangement with the landowners, these plots were spared from the insatiable teeth of the chainsaws. Once we established the plots, specialists in ecology and taxonomy began conducting inventories to discover which species were present in what numbers in the virgin forest. A year or more later, the ranchers felled the surrounding forests, turning our study plots into rain forest islands in a sea of charred trunks.

Subsequent studies identified the species most vulnerable to the fragmentation of their rain forest habitat. Among the many groups of plants and animals studied, snakes have not played a major role scientifically, but they have provided us with a bumper harvest of harrowing adventures.

The good news, to those moved and fascinated by the diversity of life on our planet, is that two hundred and some odd species of snakes inhabit the rain forests of the Amazon. This richness is of course most exciting to herpetologists, a breed of scientists and natural historians who, for some peculiar reason probably dating from early childhood, have decided to make (or is it risk?) a living catching and studying reptiles and amphibians. The bad news is that among these species is no small number that have evolved particularly nasty venoms with which to dispatch their prey. Once so armed, of course, the snakes can use this venom for defense as well as prey capture. Hence, their danger to humans, who are far too large to be considered potential prey to any but the largest of anacondas, but do, in the eyes of the snakes, pose a serious threat to the snakes' survival.

~

Manaus, in the heart of the Brazilian Amazon, is noteworthy among other things for its cornucopia of coral snakes. As many as seventeen species of these lethal reptiles call the forests around Manaus home. As if this wasn't enough, sharing this venomous stage with the coral snakes are bushmasters and fer-de-lances. These pit vipers have large poison glands behind their eyes, which give their skulls the characteristic triangular shape that warns the well-informed that this species is a touch-me-not. Unfortunately, the bushmasters and fer-de-lances have dispensed with the common courtesy exhibited by their close relatives, the rattlesnakes, of rattling to warn potential enemies of their presence. Through this most civilized arrangement, the snake doesn't waste any of its venom, and the intruder doesn't die. The lack of an early warning system, combined with nearly perfect camouflage and an especially nasty brew of both neuro- and hemotoxins make these critters among the most dangerous in the forest.

Despite the dangers, it's actually remarkably hard for anyone but a pro to find a snake in the forest. Coral snakes are mostly fossorial, living below ground or at least in the leaf litter, and rarely see the light of day, and bushmasters are not aggressive. The biggest bushmaster I ever saw, stretching out at a bloodcurdling seven feet, was caught by the fourth person in a line of four walking down one of our survey trails. The first three had stepped within inches of the beast, which was perfectly happy to live and let live. There's an old bit of jungle lore, told usually by the second of three people walking down a trail, that the first person in line wakes up the bushmaster, the second makes him mad, and the third takes the venom. In fact, if no one steps right on one of these snakes, a line of twenty could probably stroll safely past a resting bushmaster. Well, maybe ten. Everybody's got their limits.

During a decade in the forest, the project's researchers and field hands have had many close encounters with snakes of the nasty, venomous sort. Most of these stories revolve around an intrepid and irrepressible Canadian herpetologist, Barbara Zimmerman.

~

Barbara had come to the Amazon to work with manatees in 1976 and eventually earned herself a Ph.D. documenting the distribution and densities of all the populations of frog species in our forest study areas. This she reasoned was necessary if we were ever to understand what happened to the frogs in small, isolated forest reserves.

Although she specialized in the study of frogs, almost by default Barbara became the Project's resident authority on snakes. She was, however, frustrated by the dearth of snakes in the forest. At the very low rate with which she was encountering them on her nightly perambulations of the forest in search of frogs, she was likely to live to a ripe old age but not get to know much about the snakes of the central Amazonian rain forests in the process.

As Barb watched the ranchers' crews felling forest to create pastures, she saw them move through almost every square yard of the forest. She realized that we had an excellent opportunity to turn them into a herpetological army, whose marching orders were to send us, live, all snakes encountered in the hundreds of acres of forest they were felling. Their reward for each snake was pretty close to a day's salary—double that for poisonous species.

When ranchers select an area to be turned into pasture, they send in crews armed with machetes and axes to clear the undergrowth, turning the once dense and dark understory into a park-like glade. A chain-sawyer follows, cutting halfway through the trunks of fifteen to twenty trees. Finally, he fells a towering rain forest giant into the half-cut trees, bringing them thundering to the ground like an enormous set of dominoes. The danger is immense; a strong wind can set the trees toppling before the workers can get out of the way. One worker has been killed and several seriously injured on the ranches since we began the project.

Wages are minimal and all food, gas, and oil for the chainsaws and repairs have to be purchased through the company store at such exorbitant rates that a worker may often sweat his way through six months of extremely dangerous work only to find that his salary just pays his bills to the bosses. Given the minimal wages for such dangerous work, our bounty scheme drew a slow but steady trickle of snake-filled gunnysacks to our field camps and the project headquarters in Manaus. One of the first delivered was handed to Barbara by Manoel Rego, an itinerant chain-sawyer working on the ranch. As he delivered the bag to Barb, he announced that it was a bushmaster and that it had bitten him. Barb's first thought was that she was about to see her first corpse up close and braced herself to catch the poor soul when he dropped. To her surprise, he said he was feeling no pain and then added that it had been about an hour since he was bitten.

A bushmaster's poison glands are surrounded by muscles and drained by the two (or more) hinged fangs, which are the primordial hypodermic needles. The muscles around the glands are under voluntary control, so that if a snake should strike to defend itself and not to kill something to eat, it has the option of not contracting the muscles around the poison glands. In effect it does not push the plunger of its hypodermic, and therefore doesn't waste its venom. Eighty percent of rattlesnake bites reported in the United States are such "dry strikes." We suppose Manoel survived either because the snake didn't want to inject any venom or because it couldn't see him through the bag so it didn't know when to contract the muscles around the glands to inject its poison.

A few weeks later, Barbara just barely missed a flight to the next world when she inspected the contents of a gunnysack that had just arrived from the forest. Opening the bag, she reached in and pulled out a large snake, somewhere between four and five feet long. Her grip on the snake was about at what we non-herpetologists might inaccurately call its waist. After taking a good look at it, she put it back in the bag and went over to consult her snake book, as this was a species she hadn't seen yet.

After puzzling over the Portuguese description of various possibilities for a few moments, she asked me the meaning of "triads of black, white, black, white, red." I explained, and after a moment's thought she made her identification. There passed over Barbara's face that far away look that comes upon someone who realizes that she just knocked on death's door and found no one home. The first words she uttered will not be repeated here, this being a book for the whole family, but it did lead one to believe that, given the opportunity to relive the previous five minutes, she certainly would have held that coral snake a lot closer to its head.

Our snakes were always transported in burlap bags. The moments that snakes choose to effect their escape from such ignominious confinement are often not the most opportune. One incident that is easy to call hilarious in retrospect occurred when a coral snake got out of its bag in the truck coming back into town. Paulo, piloting the four-wheel drive Toyota through a ranch pasture, happened to glance down at his feet to see the brightly striped serpent slithering across the brake and clutch pedals. He colorfully expressed his displeasure at the unexpected passenger and pulled his feet away from the snake and at the same time from the brake pedal. The two passengers riding in the front seat launched themselves into the laps of those in the back seat as the car careened down the pasture road, headed straight for a fence. Fortunately for all involved, the snake decided to move on in time for Paulo to get his feet back on the brakes and pull the car to a halt just short of the fence.

Snakes, it turns out, could have taught Houdini a few tricks on getting out of securely tied bags. Two corals are still missing in action in my bedroom, which, because it was the only locked room in our two dormitory-style houses, was often the safest place, for everyone but me, to keep these critters until their final disposition was resolved. It appears the snakes found it as easy to get out of my room as they did getting out of the bag, since they have never been recovered and were first reported missing several years ago.

For a few memorable weeks, I shared living space with a medium-sized bushmaster, tightly, so I was promised, secured in an old rice sack. Some time prior, Tom Lovejoy had been visiting the project and was concerned to find a poisonous snake in an unlocked and unlabeled box in the back yard. Worried about the number of people who might unknowingly open the box and take an undeserved dose of poison, he laid down the law: All poisonous snakes had to be kept in locked boxes. Good idea. We immediately built a bunch of lockable boxes and purchased padlocks. Murphy's Law seems to be the only one that is followed to the letter in Amazonia; within about ten days we had lost the keys, or the locks, or both to all our lockable boxes.

Liberally interpreting the letter of Tom's dictum, Barbara decided that my bedroom constituted a locked box and adhered to the spirit of the law. She was, I think, kind enough to warn me that she had left the beast on top of my wardrobe. Only I and the maid had keys to the room. As the depth of the dust atop the wardrobe was measured in inches, one could safely assume that the maid hadn't seen the area in about six years and had little intention of exploring that particular expanse of *terra incognita*. I agreed with Barb that this was the safest place to stash the serpent. And there he stayed while we got the paperwork ready to ship him off to Butantan, one of the world's principal institutes for developing antivenin.

Several days later, having retreated to my bathroom for a moment's relief from the hustle and bustle of the day's activity, I picked up the novel with which I was distracting myself at the time and sat down in peace. An instant later, out of the corner of my eye I noticed something moving in the bidet, some six inches from my very exposed flank. The brain can function at speeds that make a Cray supercomputer look like a six-year-old learning long division. The instant I saw motion, I knew exactly what was moving. I'm not sure how high I jumped, but it was enough that I could look back down from somewhere near the apex of my trajectory and make sure that my reptilian roommate was still on the right side of the bag.

At my first opportunity I queried Barb as to just what it was that was wriggling around in the bidet, knowing what the answer would be.

"Oh, yeah, that's the bushmaster. Sorry. Forgot to tell you about that." She had been concerned that the snake was getting dehydrated so had taken the bag, snake still inside, into the shower to wet it down. As it seemed a bit too wet to return to its penthouse suite atop my wardrobe, she had left it in the bidet to dry out a bit and then forgotten about it.

Just what it was about my wardrobe that made it so attractive a resting place for Barbara's snakes I have yet to discover. Shortly after I encountered the unexpected visitor in the bidet, Barbara arrived in the office to inform us that someone was at the gate with a seven-foot boa constrictor, which he had purportedly rescued from some dastardly villains intent on doing the snake in. Barb asked if she could pay the 20,000 cruzeiros (US $5) requested as just reward for such an altruistic act. I thought it an unfortunate precedent to set, and with visions of a block-long line of people offering to sell us animals they had rescued from similar bands of evildoers, I offered my opinion that we shouldn't pay anything for the snake.

About five minutes later Barb returned to the office to announce that she had the snake—making no mention of the ransom—and said she had already placed it on top of my wardrobe.

I'm happy to report that despite the risks, in twelve years of intensive work in the forest, we've only had two snakebites where venom was injected—our closest brushes with disaster. One case involved a bushmaster, which apparently did not inject a very strong dose of venom. The other victim was bitten by a coral snake, and therein lies our final tale.

It was a dark and stormy night. *(Honest.)* Barbara was out on a nocturnal survey with Juruna, her newly hired woodsman, or *mateiro* in Portuguese, and Leo, a more seasoned veteran of nighttime frog and snake hunts. At about ten o'clock the light from Barb's headlamp fell upon a coral snake coiled by the side of the trail. Before she could say a word, Juruna, in a very unsuccessful attempt to please his new boss, grabbed the snake, practically trampling Barb to get to it. The snake immediately bit him on the thumb. He pulled the snake off his thumb in an instant. Barbara told Leo, who had a dull knife with him, to make an incision at the site of the bite so that Juruna could suck some venom out.

Coral snakes are rear-fanged, which means they don't inject venom through hypodermic-like fangs as do bushmasters; they need to chew on the victim while their extremely venomous neurotoxin trickles down grooves on specially modified teeth in the rear of their mouths. What may have saved Juruna was the speed with which he pulled the snake off. He immediately felt pain at the site of the bite, so there was no question that he had received a dose of venom. With a strong dose, a person can go into a coma in less than an hour and die shortly thereafter.

Barbara led her assistants back to camp, an hour's hike through the forest, where Liz Kramer, a field intern on the botany project, was sleeping. Our vehicle, unfortunately, was out of camp for the night. Leaving Juruna in Liz's care, Barb and Leo proceeded to run to the ranch's headquarters, about seven miles away, most of it over muddy roads. After a rain, gumbo-like mud clings to boots, making the going very heavy. Barb and Leo settled into a pace of jogging a mile and then walking the next until they reached the house of Omar, the ranch foreman, at about midnight.

By this time, Juruna's thumb was numb and he was in pain up to his shoulder. Omar drove Barb and Leo back to the camp, picked up Juruna and dropped them off at the ranch's headquarters. They had to find another vehicle and driver because Omar's truck was not registered and couldn't go through the police checkpoint at the edge of town. At two A.M. Barb and Juruna began a harrowing, high-speed race to town over slick, muddy roads with a long-haired wild man, as she put it, at the wheel of one of the ranch trucks. Mid-way through the trip Barb decided that the cure for the coral snake bite was going to be irrelevant since they weren't going to arrive in Manaus alive anyway.

By what may have been the second miracle of the night, they made it to town and were dropped off near the bus station. It was then three A.M. and they were at the side of the nearly deserted road, covered with mud and unable to convince a passing cab to stop for them. Eventually they made it to the Hospital of Tropical Medicine, where the staff refused to believe that Juruna had been bitten by a coral snake and did nothing for him. The principal argument against it being a coral snake bite, in the minds of the staff, was the very fact that Juruna was still alive. It was now around four A.M. and Juruna wasn't getting any worse. Barb was too exhausted to try to convince the staff she knew what she was talking about, so she decided to take Juruna home and hope for the best.

I walked into the house at 5:45 as the sky was lightening in the east, having sampled a full evening of Manaus's night life, such as it is, and found Barb sitting in the living room. My brain wasn't functioning at maximum capacity given the hour and the titer of various distillates coursing through my veins, but I was quick to realize that something was amiss. I could usually get into the house on a Sunday morning without running into a greeting party. Barbara had just gone out to the forest two days before and had not, to the best of my knowledge, been on any of our trucks coming in from the ranches since then.

Barb explained what had happened and we spent an anxious day watching Juruna, who seemed to improve steadily. Monday morning he was worsening somewhat, so I resolved to take him back to the hospital where I would insist that he be treated.

I spoke over the phone to a doctor at the Butantan Institute who was very helpful once we got around the same problem that Barb had at the hospital; someone really bitten by a coral snake is not supposed to be alive a day and a half later. After I convinced the good doctor that we really had a coral snake victim on our hands, he counseled us to take Juruna to the hospital where he should be closely observed for any signs of renal failure while our drivers were sent out to search the camps for a supply of the appropriate antivenin.

I took Juruna to the hospital, where I succeeded in convincing a nurse to accept our word and admit the patient. I said we were looking for the antivenin at our camps and should have it in town by mid-afternoon.

As the day progressed, Juruna continued to improve. The doctor in São Paulo suggested that there had not been a lethal dose injected so we shouldn't try the antivenin, even if we found it.

The story ended happily enough, as Juruna recovered fully and is still working for Barbara. He is now our resident authority on the identity of at least one species of coral snake. And he continues to lead a charmed life. Last year he came upon a fallen log across one of our survey trails. As he went to cross it, a huge bushmaster, which was coiled on the other side of the trail, struck at him. The snake's fangs would have plunged straight into his chest—where do you tie the tourniquet?—had there not been a sapling between the two of them. The snake struck the half-inch diameter tree just inches short of Juruna's chest and fell back into a coil as Juruna beat a hasty retreat.

The Ghost in the Machine

by Phyllis Lee

For people doing research in exotic areas (cold, hot, distant, wet, gloomy, or green) one piece of equipment is essential—their instrument of transport. I suppose that most field biologists would instantly look at the soles of their feet, count blisters and tick bites, and think of tramping through thick, endless vegetation. For others, however, some alternate form of transport (a boat laboriously paddled, the back of an elephant, even horses) comes to mind. For some, airplanes or even helicopters get the researcher, baggage and all, to the site and out again. But for biologists studying the large, charismatic mammals of the great National Parks of the African savannah, transport often takes the conventional form of a car. But a car is never just a car; it is an animate and often malicious being, nevertheless critical to the success of the project.

Why does this foul, polluting object loom so large in one's affections? The car serves as the primary tool for getting in and, often more importantly, out of a study area. Whether a small or large animal watcher, a car is the main way of finding one's subject of study. It is a metal cocoon providing protection from harsh ecosystems where exotic subjects live, saving one's feet and, occasionally, life. It is the continuing link with the "outside world," where things like engines, refrigerators, and computers work with some degree of regularity; where petrol and tires can be bought and spare parts found; where water comes from taps and is (usually) drinkable; and where the beer is cold. It is a psychological escape from the day to day difficulties of the field. When things get hard, one can hop into the car and flee—or at least, dream about fleeing. Every time the ignition is switched on, there is a momentary leap of the heart. Will it start? If it doesn't what are the alternatives? (Usually none but a day without work, filled with greasy, smashed fingers, and frayed tempers.) In the field, the car should be a reminder of order amongst chaos, but it is usually in a state of chaos itself.

Why a car, some intrepid foot-bound biologists ask? In some national parks, quite rightly, you need to receive special permission in order to wander about on foot. After all, it is hard to keep up with hunting lions on foot in the dark, and not terribly safe either. Nor is it great fun to observe elephants feeding when the next bush they eat may be the one you're lurking behind. And many smaller mammals inconveniently live some way from one's residence (be it tent or hut), thus the pressing need to find them before they disappear on their daily pursuit of food requires the early morning "game-watcher's commute." The car then becomes the center of the universe.

My first introduction to serious fieldwork with African mammals was on foot, treading through the damp, snake-infested, insect-laden, green hell of a semi-deciduous rain forest in western Tanzania on the shores of Lake Tanganyika. Sooner than desired, through a variety of unusual circumstances, I abandoned this site with its reliance on boats and my all too soft, unhabituated feet, and moved inland to an ancestral landscape where images of hominids still wandered freely through the imagination. I upgraded, in transport terms, to the back of the most robust Chinese-built bicycle I have ever encountered. We had two bicycles among four people—two people to pedal and two to sit on the back going along the flat or downhill. I made sure I was a sitter, not a pedaller. There is little to compare with the thrill of speeding down a steep, bumpy, dirt track, as I did one day, in search of a troop of baboons on the back of a bicycle, with a grazing, looming elephant at the bottom of the hill and 130 pounds on the back of the bike giving it that critical extra acceleration. Fortunately the bike came equipped with a bell, which repelled the elephant with ridiculous ease. It was my first encounter with an elephant and in hindsight, I would never attempt it again knowing what I now do about elephants and the speed of bicycles. Who needs a car, I thought, until pushing said bike up the hill after a long day's trek after monkeys. Further encounters with interesting species, a pack of curious wild dogs in hot pursuit of some equally hot hominids, several hours up a (small) tree with a couple of hundred elephants eyeing our tree with greedy eyes and tusks, and the odd rhino and buffalo taking umbrage at the intrusion of flat-footed researchers into their domain convinced us of the necessity of a car.

Into our lives rolled an ancient Land Rover; a dilapidated, clapped out, worthless, rusting, gasping, and continually breaking-down piece of equipment. "TDU," as it was known by its license plate (instilling almost immediately that requisite anthropomorphic touch), was hardly to be blamed. It had served a long previous existence (I suspect in long distance haulage) and it had now come to retire in the luxury of a savannah national park in the company of several incompetent amateur mechanics who would tinker in frustration but couldn't make the damn thing go.

TDU did run one time. It got three of us out of the park and into the nearest large town for a supply run. This involved negotiating some of the most beautiful parts of the park on some of the worst roads, heaving the hulk across a broad (seemingly endless) river on a hand-pulled ferry/float, and then driving for many hours on a corrugated road between small villages dotted with mango trees. We reached town and passed the night in luxury (cement walls and no animal night noises—the ultimate bliss to bushed bush workers). The next day we managed to get hold of canned milk, margarine, jam, and a crate of tomato sauce. There was a package waiting at the post office, containing a bag of M&Ms (simple pleasures for deprived chocolate junkies). We found fresh cabbage, onions, green tomatoes, and a variety of unripe fruit to store and consume over the next month or so. We even filled up with petrol. In an ebullient state, the three of us started back toward the park.

A mere thirty miles or so away from town, we developed a flat tire. Fair enough, changing tires is pretty standard, even if it means unloading several months of supplies. Naturally, we greenhorns had neglected to leave the spare accessible. Only that wasn't the end of it. While getting the spare into position, an ominous hiss began to make itself heard from yet another tire. Two flat tires: one spare. We never thought we might even need *that* one. After all, there's always AAA and emergency call boxes for help!

That day we coped without the requisite pump, patch kit, and extra tubes that I now carry everywhere, even in places with "comfort stations." In our state of exuberant naiveté, we thought the solution to our problem was simple; one person hitches back into town with one tire and hitches back when it's fixed. Simple. One slight problem: there was no other vehicle in sight; none had passed the previous day and none was due. Nothing to do but wait. Late in the morning, our rescue, in the form of the park's truck also heading to town for supplies, rolled up. Tire and person embarked, leaving two of us with the car and the supplies. Did I say one slight problem? Two slight problems. None of the food was ripe enough to be edible, or worth eating straight from the can. The longed for M&Ms vanished under the stress of hunger and heat (yes they *do* melt in your hands). Then the thirst began. There was, of course, a jerry can of water, but even we weren't that stupid. The can had previously contained petrol and was seriously rusted, thus the water was undrinkable. The heat intensified, the M&Ms caused throat constriction, and the unripe bananas began to look good.

Late in the afternoon our colleague returned with the repaired tire and the most delicious, sticky, orange Fanta ever to quench a Saharan thirst. The park's driver said we should follow, so we set off gradually falling further and further behind the truck's thick cloud of dust. Dusk was approaching; the ferrymen ceased operation at sundown and unless we arrived with the park's truck it was another night on the road. TDU chose that moment to fade gently out. The engine died. There was no way to coax it back to life. Rounding up the usual suspects we found little but chronic decay. The points, plugs, and carburetor (heart, lungs, and brain) were tested and tentatively pronounced sound. There was nothing to do but wait until dawn to begin some more serious dismantling.

This time we were lucky. TDU had grounded on the outskirts of a village. There was a small stream with pools of water, and a helpful man showed us to a spot where we could get a small bowl of water each. But no food was available in the village—there was a wedding planned and everyone was going to attend. The drums for the party began with the rise of the full moon and went on all night. People wandered to and fro along the road the entire night, obviously enjoying themselves. We stayed in the car and, three to the gearshift, tried to sleep. It was a beautiful, cool, clear, moonlit night that seemed to last forever.

First thing in the morning, after seeing the sorry state of the stream in daylight, we headed for the tiny village shop. Soft drinks were sold out after the previous night's celebration, but miraculously there were a few warm beers. Consuming warm beer in the early morning after little food or sleep tends to induce a mildly euphoric state. Back at TDU, the hood went up, heads went under, and the search began. The tinkering had been underway for some hours when the park's Land Rover, with mechanic, arrived. The people on the truck realized that we hadn't managed to follow them, but could do little until daylight when one of the few vehicles was free. The mechanic put his head into the bowels of TDU and emerged after about twenty seconds with the comment "try it now." It started; one tiny loose wire had sent forty years of British engineering careening to a grinding halt. It was dusk when we arrived at the river, and it was doubtful if we could cross. Another night in misery was avoided through the kindness of the ferrymen, and we eventually arrived back at our camp.

Such was my introduction to the internal combustion machine and its use during fieldwork. We spent most of our time on foot or back on bikes, with the car up on blocks. On those rare occasions when TDU was not serenely resting on blocks and we unreasonably tried to use the damn thing, a tire was flat or some mysterious fiddling was required under the hood, or both, and tempers proved as short as TDU's wiring. The carburetor had a special gremlin and I spent many hours watching its dismantling and reassembly time after time after time. Eventually we deserted TDU altogether.

The lessons learned under TDU's hood were extremely useful for the next session of fieldwork. I bought a *brand new car.* Admittedly only an 800cc two-stroke, soft-top jeep, but it rarely failed me through any mechanical fault. It did have a penchant for attracting animals. Vervets would continuously climb through the soft-top and crap on the seat. This would only be discovered after sitting down. The dampness would slowly permeate one's consciousness. Baboons would try to steal my lunch or bounce on the trampoline-like roof. Bushbabies also seemed to use the roof to enhance their nightly territorial displays, making a surprising amount of noise for one-pound animals. But the worst was my experience with the elephant.

I was out watching monkeys and came across a large bull elephant in what Cynthia Moss and Joyce Poole, my colleagues, have described as musth—the aggressive sexual state that comes upon a bull yearly. This male resided in my monkey's area and I knew that Cynthia and Joyce would be interested in the sighting so I spent some time sitting in my car drawing his ear so that they could later identify him from his characteristic earprint. I only left when he began to become overly interested in my presence—starting to display, rushing at me, and flapping and waving his ears. The car retreated in a roar of two pistons and a cloud of oily smoke. My mistake.

The next day, peacefully tramping on foot after the monkeys some distance from the car, a large shape loomed into my peripheral vision. It was very close. I could smell the specific sexual odor of the bull and hit the ground, trusting the tales that an elephant's vision was poor if you were horizontal (harder for them to get their tusks into you, too). The bull marched past me and with a huge sigh of relief, I got up, dusted myself off, found the animal I had been observing, and carried on. Until I realized where he had gone. Without deviation or hesitation, the elephant headed straight for my poor little car. Dodging between bushes and trees I got closer. Three thoughts went through my mind. Firstly, if he destroyed the car, the insurance company would say it was an Act of God, so bye bye insurance. Secondly, I only had two weeks to go and was collecting my last crucial data. What would I do without the car. Thirdly, it was a long walk home (eleven miles) through lion, buffalo, and elephant infested terrain.

The bull charged the car, he backed off, turned sideways and pretended to eat. He tossed dirt at it, and charged again. He tusked the radiator, backed sideways, danced around, and attacked yet again. This went on for forty-five minutes with me lurking in the grass in a state of near panic about the outcome of this contest of David and Goliath. David clearly had forgotten his catapult. Finally, tossing his head, the elephant began to move off, then he turned and ran straight at the car once more. The car stood its ground and remained intact. Realizing that it wouldn't flee, and finally having met his match, the musth bull retired ungracefully from the scene. I called it a day and fled the scene with shaking feet on the clutch. I reported this misbehavior to Cynthia, and showed her the drawing of the bull's ear. She laughed.

"It's only Sleepy. He always plays with cars."

It didn't look like play to me, but then I didn't know much about elephants at that time.

The poor car was pressed into service in many other ways besides having to defend its honor against overly aggressive and sexually desperate elephants. I was always glad it was somewhere nearby, especially when the hair on the back of my neck would rise under inspection from some unseen presence. This usually turned out to be a lurking lion, gazing with disdain at a crazed pink person wandering feverishly after small monkeys.

I again relied on a car for my next field session, but a slightly upmarket jeep this time. It had a penchant for seeking out and then sinking up to its axles in the only patches of mud for miles around. Fortunately it was light enough to lift out if four or five strong men could be found. But long walks for help seemed to be the order of the day. It also developed an enmity with a musth bull elephant who would leave his female consort to challenge the jeep from several miles away. Once, while watching a fantastic fight between two musth bulls, the jeep and I suddenly found ourselves the target of both bulls' redirected ire. We fled with two enraged bulls in hot pursuit. The ground wasn't conducive to speed, but that scarcely mattered. We hit the flat and sped away. It was only when we got home that I realized that the chassis was cracked. One of the bulls was later found dead, the victim of the clash between the rival bulls.

The "two flats/one spare" syndrome was common, and even escalated after I managed to get an extra spare. One day was spent with the car up on blocks with four flat tires and only two spares operative while the punctures were being repaired. Batteries would discharge at inconvenient moments (while being chased by a bull or in the middle of sampling), oil filters would work loose, half-shafts would break, and generally all things mechanical would sooner or later develop some fault. Whenever a friend or colleague failed to return by dusk, cars immediately became suspect and the search would be launched, looking for the flashing lights of a defunct, stuck, or out-of-action car. One New Year's Eve celebration was spent searching for two friends stranded in the middle of a dense swamp.

Despite all this, the car remained essential to data collection and was also used for escape from the field. I learned a great deal about how cars are put together, and how to ensure survival when the inevitable problem would arise. (But sooner or later, as a field biologist once said, "I was down to my last croissant, so I knew I'd have to walk for it," becomes the story of all who rely on the car.) Many a weary biologist has limped home (throwing pitiful stones at predators or other beasts which pursued him or her) seeking to rescue not themselves, but their car. Of course, cars aren't the only cause of problems. There was the time that a friend dropped the keys to her airplane into the pit of a long-drop toilet.

Loved, hated, worked on, and worried about, cars sum up the existential crisis of fieldwork in the carnivore-laden African savannah. They follow the ultimate dictum that whatever can go wrong will do so. They metaphorically abandon one in hours of greatest need, and, actually, often create those desperate situations. They fall into "man-sized holes" in the road, impale themselves unaided onto logs, accelerate into the face of danger, and provide at the same time the lifeline to sanity while driving one insane. They represent all that is best and worst, unpredictable and addictive, about western ways of life. Where would we field biologists be with just our feet and brains? Like our australopithecine ancestors who roamed the savannahs, probably extinct.

An East African Survival Course

Herbert H. T. Prins

TANZANIA'S LAKE MANYARA National Park, where I worked for four years, is a gem. It spreads below the Ngorongoro Highlands and fills a part of the Great Rift Valley, tucked between the lake and the incredible steepness and breadth of the Rift's escarpment. The amount of rainfall at Manyara is low but it has many springs emerging at the foot of the escarpment, and together with its rich volcanic soil, the permanent abundant water brings life to one of the highest densities of large game in the world. Imagine fifty sturdy buffalo on every square mile and add to the same square mile fifteen impressive elephants, clownesque wildebeest, fat zebras, dreamy impalas, black rhinos, sleepy lions, towering giraffes, priggish hippos, proud leopards, quarreling baboons, and lovely bushbuck. What a world!

I studied buffalo, their behavior, social organization, and feeding ecology. However, despite their reputation for aggressiveness they never caused me any trouble—perhaps they knew that I meant business. Of course buffalo are very interesting, but they are not the focus of my story. They are a bit like cattle, and up until now I never met a farmer who told an exciting story about his cows.

When I worked on foot in the park, I always took two rangers with me: Mhoja with his old Mauser, and Renatus with his Holland & Holland, and both with only two cartridges in their pockets. Mhoja had been a ranger for over twenty years and Renatus was a veteran of the 1978 Tanzania-Uganda War. I have never seen these two men afraid, and apparently they hardly ever worried, not even when a friend and I were chased by a mad elephant herd. While we were running for our lives towards the rangers, Mhoja just calmly fired over our heads and over that of the elephant just fifteen yards behind me. When the shot cracked the elephant screamed off through the bushes. This frightening encounter unnerved me for more than a few hours.

I believe these men were fearless because of their keen knowledge of the bush. Both Mhoja and Renatus were ethnically Sukuma. When Renatus turned to a civilian life after the horrors of the war he was only twenty-two, and was more-or-less adopted by Mhoja who was his senior by more than twenty years. In addition to love and respect for nature, he taught Renatus much about carpentry. Together they made music in the camp and were great story tellers. Mhoja started out as the pupil of a witch doctor, but when he was asked after his apprenticeship to cut off one of his thumbs to show that he wanted to continue this particular career he decided that this was too hard a way to earn a degree, and he became a carpenter in the Tanganyika National Park Service. He was spotted at the end of the fifties by Desmond Vesey-FitzGerald, the nestor of East African ecology. Vesey taught him the Latin names of all plant species in Manyara. Twenty years later he still knew them all, but also the Sukuma names and many Maasai names. Time and again when we went in the bush he asked me to repeat what he knew. Sometimes he used Vesey's way of talking (very British) and even Vesey's voice: a living tape of a man long since dead. Mhoja taught me how to track animals, and constantly rehearsed the difference between prints of bush-buck and waterbuck, the smells of different dead animal species, or showed me how to follow snake tracks in the grass.

On one particular day in March it was pay day, and the park was empty; all the rangers had left the park to spend a few days in the village. My friend Paul Loth and I, however, wanted to continue working on a landscape ecological map of the park, so we decided to go out alone. We had to make a *relevé* in the Endabash area on the escarpment. A relevé is a representative plot of vegetation, varying from two square yards in simple vegetation to about a hundred square yards in a complex savannah. In such a relevé, botanists estimate the cover of all individual plant species. In that manner they get a representative description of a vegetation unit. I drove the car through the bushes on the slope, until Paul said we were in the area that he had selected from the aerial photo. The relevé plot was densely bushed with a visibility of less than five yards. We walked through several acres for about an hour, writing down all plant species and their cover, and we encountered some quite fresh rhino scrapings. Paul and I then walked back to the car to collect an auger, and returned to the center of the plot to take soil samples. We sat in the sun—it was about eleven in the morning—like two little boys playing with sand, to determine whether the soil was a clay-ey loam or a loamy clay, when suddenly I heard the breaking of a twig. It was like the sound of somebody snapping a match stick. I told Paul to move to the car but he insisted that it was only a little bird. After many months in Manyara I knew, however, that this could not be the case, so we ran for the car, leaving all equipment behind.

Paul scoffed at me because nothing happened at all. The bush was completely silent, not a bird moved or a wind stirred, but I was still uncomfortable in this peace. I put the car in four-wheel drive and drove up the slope until it became too steep. We were just twenty yards from the spot, and still nothing happened. Paul wanted to go back and continue our work but I suggested that he stay in the car, while I climbed a termite mound some eight yards to my right to scan the bushes. Just as I'd gotten out of the car, a rhino, halfway between the termite mound and the car, came charging at full speed. I was terrified and jumped behind the car, surprised at how quickly fear could move me. The rhino thundered off without turning. No little bird breaking a twig! A few minutes later we collected our gear, and discovered that the rhino had circled our site several times only yards away from where we were. We had not heard a sound till the breaking of the twig! Apparently, the rhino wanted to find out what we were doing in its territory. It's impressive how these inquisitive creatures can slide their tank-like bodies silently through the bush.

Another day we were looking for a buffalo calf. The day before I had seen it on the mud flats near the lake, apparently deserted by its mother, or perhaps the mother had been killed by lions. Lions depend to a large extent on buffalo in Manyara—nearly ninety percent of all buffalo ultimately leave this earth through the mouth of a lion. Because of the high density of large game in this park, there are also many lions. They still, occasionally, kill people too. As I wanted to locate the presumably dead mother or the calf, I sat on the roof-rack of the car, while Paul drove. We cruised through the bushes and because I sat so high, I had a rather good view. After searching for an hour or so finding neither calf nor cow, we decided to search in the adjoining Acacia woodlands.

I searched intently between the shrubs and did not pay much attention to my surroundings. Paul enjoyed the driving, listening to Rossini. A few hundred yards away a herd of elephants quietly browsed, and there were no tsetse to bother us. I felt at peace with the wilderness—a seeming paradise. Luckily our ancestors were no fools—our existence proves that they survived long enough to beget a next generation—and I think that I inherited from them my sixth sense, well-known but elusive to science. Without warning I got the feeling that there was something wrong. Terribly wrong. Immediately I spotted them, a pride of lions in the Acacia tree just in front of us, less than ten yards away. There I was sitting on the roof of the car, while Paul was driving slowly in the direction of the tree, lost in his Rossini reverie. Perhaps it sounds strange, but I had no time to become afraid. I shouted very softly to Paul—so that he would hear it above the music but without focusing the lions' attention on me—to drive full speed. Paul reacted instantaneously. The car sped forward, and I laid flat on my back on the roof-rack. I will never forget the sight! We drove straight ahead under a large lioness, her claws only inches from my stomach and then my face. She looked straight at me with her cold yellow eyes.

As can be understood by now, I don't particularly like lions, especially when they are roaring outside the camp where I slept with only mosquito screening between myself and the outside world, or when they walk through the camp to dig up and destroy the water pipe, causing days of restoration. Yet, their beauty makes me forget how dangerous they can be. One day, I came to my camp at Ndala River quite early in the afternoon. Ndala Camp is finely situated at the foot of the escarpment along a waterfall. The site was chosen by Desmond Vesey-FitzGerald together with Iain Douglas-Hamilton, the elephant expert. Later, John Scherlis and I restored and improved it. Apart from being a gifted elephant observer, John was a genius when it came to designing and building. We had,

I think, the most elegant toilet of all camps, with a Dutch barn door. Many bird observations were made from this toilet, resulting even in a scientific publication. The toilet, however, was not mentioned under that obligate section called "Methods." Apart from the toilet, we constructed a great shower. Just a bucket hanging from a tree in which Sjabani, the servant, every evening poured the hot water, as a loving father, at the perfect temperature. It was a pleasure to stand on the smooth river stones from which we had made the floor beneath the shower-tree, albeit sometimes eyed by our wild camp leopard.

Near the camp a waterfall spilled into the riverbed where there were many large boulders and a beautiful pool. At this pool, bushbuck, impala, buffalo, and baboons come to drink. During the dry season, one could see here scores of elephants. They even entered the pool to swim around and play in it. (Today the elephants have all been killed and the pool is gone.) One day when I returned early from the field, I decided to have tea served outside. At that spot, the river bank is several yards high, overlooking the pool and the river without disturbing the animals. We have even thrown pebbles on the back of an unsuspecting elephant without disturbing it. I drank my tea and contemplated life when two large, black-maned, male lions strolled to the pool for a drink. They were so close that I heard them lapping the water. After satisfying their thirst, they crouched behind the boulders to start their vigilance. I called Mhoja and Renatus, Sjabani, and the cook, Tseama, and we sat a few yards above the lions waiting to see what would happen. Nothing. No animal came so I went inside to have dinner. After dinner—it was full moon—we went to the same spot again. The lions were still crouching behind the boulders. The air was heavy with smells of the tropical night, the bush was silent, the lions were silent, and so were we. Then out of the darkness of the trees a group of elephants moved as ghosts along the river. In the lead was a small calf of about three years, followed by its mother, and in the center of the group a large matriarch towered above the others. The lions flipped their tails and slowly the elephants moved in our direction, heading for the pool. When the baby elephant was close, the lions attacked. In a flash, the matriarch rushed toward the lions. Everything happened without a sound, making the scene quite eerie. The lions braked, turned tail, and sped in our direction. They jumped up the river bank and ran between our little group and the hut. We sat frozen but the elephants moved on as if nothing had happened. They just continued their walk to the pool and started drinking. This, apparently, is the way to behave in the African wilderness: always be on the alert, stay calm, never panic, and continue the work you came for. I was learning this but only after being tested several more times.

One October day, during another hot and dry season, I drove by car to the Endabash area. The last several weeks had seen a huge influx of vultures in the park, and I then discovered that many impalas and elephants were dying of the dreaded anthrax disease. I encountered several fresh carcasses, but further south the disease had not taken its toll yet (unfortunately at the end of the epidemic about ninety percent of all impalas were dead). I never liked the Endabash area because the elephants could be very mean there, attacking without provocation. Cautiously I searched but did not spot any, so I drove my car through the dense bush to an area where buffalo roamed on the lush swampy vegetation. I conducted my telescope observations for about an hour when, some sixty yards away, my eye caught sight of a dead rhino carcass. Peering through the telescope, I saw to my dismay that the horns were gone from this very big bull, apparently killed by poachers. Sometimes the poachers were Maasai who killed them in exchange for a beautiful living zebu bull which they got from the criminal middlemen. One time, poachers had even used semi-automatic rifles; but the Iraqw hill farmers used a different technique: A group of men sat on the escarpment searching for a sleeping rhino somewhere below them. Then they would climb down and set up an ambush close to the rhino. The quickest runner was selected to sneak up on the sleeping rhino; he would wake it and let the rhino chase him at full speed into an ambush. There some ten to fifteen men were ready to kill the rhino with their razor-sharp spears.

The fact that the present dead rhino was still bleeding, scared me. I felt the brooding eyes of many a poacher crawling over my body, and I imagined spears or rifles aimed at me. Being totally alone made me feel vulnerable, and I knew it would take days before anybody would miss me. I quickly started the engine, rushed away, and drove twenty-four miles north to find the Warden of the park, Elias ole Kapolondo. When I entered his office and told him what I had seen, he jumped up and said he would collect a team of rangers to search the area. A few minutes later he returned a bit sheepishly to report that all his rangers were apparently in the village, drinking beer and charming their wives because it was pay day. So, Kapolondo decided, we had to go alone. Off he went to collect his rifle from the armory, only to return a few minutes later to inform me that the junior Warden had taken it. But this was no problem—we then would go without a gun! Now Kapolondo had been in charge of a section of the Serengeti National Park, and he had set a record, dubious perhaps, but much applauded, arresting over 700 poachers in one year; and around Manyara he had organized a very effective informers' network. As a result, the few poachers he caught in the park were all brought to court within a year. His informers' network, although not very sympathetic, proved later to be much more effective than a lot of patrolling. Kapolondo was a Maasai and totally sure of himself as an embodiment of the Law. We drove the twenty-four miles back and relocated the rhino carcass. Sure enough, there were no horns, and there were a few holes in the skin. The dead animal had started bloating already—it was actually bubbling and boiling because of its mass and the heat of the full sun—and so we could not see what type of holes these were. Nevertheless, we were both quite sure it had been killed by poachers, and decided to track them: Kapolondo and I, walking side by side, bent over with hands on the back and heads bowed down. The bush was silent and hot.

Suddenly we heard an "oomph" close by, and then another one. Although the sound was vaguely familiar, both Kapolondo and I turned around and to our utter amazement we saw two lionesses standing on the ground a few yards away from us and four others still in a small Balanites tree we just had walked beneath. At that moment another lioness jumped out of the tree like a big lazy pussycat and when her forelegs hit the ground, we heard "oomph" again. It was the sound of the air being expelled from her lungs. When the fourth lioness then jumped out of the tree, Kapolondo just remarked, "Oh, lions." The only thought occurring to me was that this was extremely clear-sighted of him, and I answered, "Yes, lions." I don't remember being afraid at the time—it happened too suddenly—because Kapolondo reacted so calmly, or more accurately, he did not visibly react at all. So when he suggested we continue our search for the poachers, I just joined him: heads down, hands on the back, gazing at the trail. Out of the corner of my eyes—because I was not searching intently for tracks—I saw the lions following ten yards behind us.

After a few minutes they stopped, lying down on the trail, while we continued our search. However, we found no tracks, so Kapolondo suggested we go back to the car. We walked back along the same elephant trail and to my horror I saw the lions still lying and sitting there. Kapolondo plucked a blade of grass and stuck it between his lips. I thought "Okay, if that is the way to approach five lions, I will do the same." I still don't understand what happened next, as we walked slowly towards those large predators with their man-eating reputation. The lions leisurely raised themselves; two went to the right of the trail and three to the left, and Kapolondo and I walked in between, like the Children of Israel crossing the Red Sea. After passing, the lions followed us again back to the car, and when we approached the Balanites tree again, the last one jumped out to join our small circus. Certainly I was relieved to be back in the good old Land Rover, and asked Kapolondo whether he had been afraid seeing these lions so close by. He smiled, and said, "Of course not, I am a Maasai, and we Maasai don't have to be afraid of lions." Later I discovered that a blade of grass can be used as a signal that a Maasai wants peace but it can also be used to curse somebody. Did the lions know?

Kapolondo, however, still wanted to know what had killed the rhino; as I said, he was a thorough and committed Warden. We then collected some rangers from an outlying guard post who were skilled trackers. They discovered that shortly after the rhino had died, the horns had fallen off—rhino horns are only attached to the skin, not to a bone—and hyenas had carried them away and dropped them somewhere in the bush. They thought it unlikely that the rhino had been killed by poachers because they found the two horns. The task rested on me to find out what had killed this big bull. So I ordered everybody to stand above the wind, cut off an ear, took a blood sample, and looked through the microscope. Then I saw the real killer of one of the last of Manyara's rhinos: anthrax.

Nature was teaching us lessons about how to behave in the bush under stress. It also taught that the outcome of many an encounter with the forces of nature is unpredictable, and that "being well-prepared" often does not mean very much. Even if Kapolondo had taken a gun, would it have helped if six lions had decided to jump on top of us?

This reminds me of one final encounter I had with a big cat. Ndala Camp was situated within the Park. The camp was the feeding ground of elephants, baboons, dikdik, etc. It was also used by lions and leopards. I daily recorded the tracks of leopards for nearly two years, and on average every three days we had a leopard in the camp. The first time I saw one was when I came back from the Serengeti in the middle of the night together with another colleague. We unloaded the car, with two hurricane lamps standing at both sides. When we were ready, I went to brush my teeth, and discovered a leopard sitting on its haunches behind a bush, just three yards from the car. One night I even woke up—the sixth sense again—because a leopard was watching me through the mosquito screen, with his forepaws resting on the window sill. This unnerved me, because in the village above the escarpment several people had been killed by leopards. Apparently they went after the goats people kept in their huts at night—when the leopard jumped through the thatched roof into a dark hut it sometimes accidentally killed a human instead of a goat. But when Sjabani returned home one day, and told that he had killed an attacking leopard, I started doing just what Sjabani and Tseama the cook did: at night I walked around with a spear.

One night I was totally alone in the camp. I had gone to bed early because I felt a bit lonely, and thus unsafe. After reading a few articles about ecology by the light of a hurricane lamp, I wanted to sleep. The bush was silent and very light because of the full moon. Apparently, no lions or leopards were close because the baboons and vervet monkeys sleeping on the cliffs behind the camp were silent. This was one of the lessons I had learned: listen to other animals and their warnings. I slept soundly until about two o'clock in the morning, when I woke up because it was hot and I had to pee. I was uncomfortable with the many animals in the camp at night, and did not want to go to our fine outhouse with its Dutch barn-door, fifty yards through dense bush. I had stopped doing that months earlier, as had my guests and students. I went outside, and stood naked on the stoop of my hut, looking at the silent bush bathing in the moonlight. I felt happy, and heard some elephants playing in the pool below my hut. Far off was the sound of a pygmy owlet. I started urinating from the stoop, when from around the hut behind me there came softly walking a male leopard. I saw him when he was less than three yards away and, by reflex, I froze in the middle of my act. He passed by, a mere two yards away, and looked at me without showing any interest and without interrupting his gliding, soundless walk. After passing, at some six or seven yards, he suddenly jumped more than two yards in the air, and caught a flying bat between his claws. Again without interrupting his walk, he looked back at me over his shoulder with the bat hanging between his teeth. I saw him smile. It was as if he said, "I could have gotten you without any effort, but why bother? We both enjoy the soft night and the beauty of life in the bush." I went back to bed, and was asleep within a few minutes.

Wildlife in Kilgoris Hospital

by Pieter van den Hombergh

KILGORIS WAS REMOTE after all, I realized, as our driver took a left turn from the tarmac onto a slippery dirt road near Keroka. This road would lead to Kilgoris, a tiny district town with a market, a couple of bars, and some houses. The hospital, named St. Joseph's Hospital, was a mile away hidden in trees on a little hill. This was the place to which we had been assigned for a full three years. The hospital had been set up to provide health care for the nomadic Maasai herdsmen living on the fringes of the Maasai Mara Game Reserve. Around this wildlife sanctuary is a large conservation area patrolled by game rangers, in which the Maasai, cattle, and wildlife coexist, albeit somewhat uneasily. I had dreamed of working in this area since I started medical school and even before.

The landscape changed to grassland with dots of bush evenly scattered. It was an unexpected but very welcome sight after leaving overpopulated "Kisii-land" with its stamp-like appearance of meticulously-trimmed plots. Coming from a densely-populated country myself, I savored the sense of space and the scent of the wild. The three of us, my wife, my son and I, had set out a few days earlier for, what had been called in the job description, a very "desolate and inaccessible" place some thirty miles from the escarpment of the Maasai Mara Game Reserve; now we were arriving there.

"You are going to have a tough time," said Father Priems, director of the agency that interviewed me for this job, referring to the principal inhabitants: the wildlife. I had difficulties hiding my joy and I carefully suggested that it could perhaps suit us. It would.

It was evening by the time we reached Kilgoris, and we saw an increasing number of fluorescent eyes reflecting our headlights, further confirming my expectations of wildlife in the area.

"No, no," said Sister Marie-Johanni who greeted us on our arrival, "those were cattle," and she was probably right.

We were welcomed by our future colleague Jacobus Bakkers. A little bewildered, we lingered in the small porch. Standing at a mere six feet, six inches I had to circumvent a buffalo-head trophy, hanging just next to the entrance. Jacobus figured it would be an appropriate introduction to the area by telling us its story before inviting us in. Just a year ago the buffalo had run into a snare set by poachers. The snare cut through his right eye and into the bone of his nose and brow. The buffalo broke loose and became excessively bad tempered: he gored two villagers into the hospital before being killed by a ranger. Standing on the porch on that first dark night, I didn't know what to think of such a story. Jacobus saved us further details, but traumatology has come to mean something different to me ever since.

Injuries caused by game animals are only one, and not the greatest, cause of accidents in the Kilgoris area. Indeed, the category in the hospital's yearly report, "injury inflicted by buffalo and others," was no larger than other kinds of accidents, and much less than, for example, the numbers admitted for child birth. Nevertheless, this relatively small category captures much of the hardship of life in remote areas.

My first glimpse of the potential hazard of living surrounded by wild animals occurred when I approached my first psychotic patient, or so I thought, in Kilgoris hospital. She had come in at night very upset and paranoid, but not aggressive, so I tried a "human" approach. An arm on her shoulder and a face to face gaze, standard techniques in modern psychiatry in the early seventies, nearly proved to be a fatal mistake. As the patient fought her craving to bite me, her eyes pleaded with me to go away. Her incapacity to swallow water (erroneously called hydrophobia at first) confirmed my supposition that she had rabies. Rabies is carried by dogs who have been bitten by infected jackals, hyenas, and mongooses. Patients with rabies are themselves bewildered by their uncontrollable symptoms, which they experience with full consciousness. Over twenty people died of rabies alone in some years, and this did not include those who died at home. There was no hope for her, so we watched her at some distance. She had impulsive outbursts of dog-like behavior, confirming reports I had once read that the victim adopts the behavior of his or her deadly vector.

Since everything was so new during the first weeks, I didn't realize what effect this event had on me. Educated in Holland, surrounded by the built-in safeguards against professional hazards, my routine excluded primary responses like fear or disgust. Yet in Kilgoris things were different. I was responsible for patients who lay only a stone's throw from my house. I could no longer take for granted the sterile protection of the hospital. After the inevitable exhaustion of the day, I would normally fall asleep with no problem. But the nightmares lasted till dawn. In those dreams I experienced all the agony that one bite caused in the days before Pasteur conducted his experiments.

Soon after I had settled in, Jacobus informed me that he would like to take a week's holiday. I wasn't looking forward to the idea of running the hospital alone. Jacobus planned to camp with the Maasai during the period of their *Eunoto.* This is a very special ceremony marking the transition of warriors into elderhood, occurring about once every seven years. Prior to the initiation of the ceremony a lion has to be killed, and its skin wrapped partially around the kudu-horn that announces the start of the ceremony. To my great relief the date was postponed. The lion was the problem. Lions were not to be killed, and the government, who didn't like this "Maasai nonsense" at all, could arrest a lion-killer.

Such restrictions complicated our work in the hospital. For example, one morning a man was brought into the hospital with his arm twice the normal size. In defending his cattle against a lion he had been deeply clawed. Pus was oozing out of puncture holes that were inches apart, suggesting an enormous claw. The injured man had killed the lion but was afraid of getting caught by the authorities. Desperately fearing the repercussions, he resorted to hiding for several days before finally deciding to seek help in the hospital. His delay nearly cost him his arm, but because the bone was not infected amputation could be avoided.

During the preparation for the Eunoto, we went to visit the site. It was the most colorful and impressive picture one can imagine with every Maasai in ceremonial outfit, constructing the ritual *manyatta* (kraal) designed to accommodate the Eunoto. We asked the *moran* (warriors) if they should be hunting a lion. They laughed, appreciating our somewhat teasing recognition of their bravery.

At Jacobus's request one moran came to show the scars in his neck. He had been admitted to the hospital some months ago, being attacked by a lion. Before his fellows could interfere the lion had cut his throat. He had to breathe through the opening in his neck, but made it, nevertheless, to the hospital. His recovery certainly added to the legitimacy of our presence in the eyes of the Maasai.

Lions, however, actually caused little trouble in the years that I worked at the hospital. It may seem strange in an area about one quarter the size of Holland, with a population of 100,000 people, and large numbers of lions living on the borders of the game reserve that so few lion victims ended up on our operating table. Are lions not so aggressive, or is every assault fatal? Dead people are never brought to the hospital. Perhaps the spear-wielding Maasai are too daunting a prey? Perhaps the famous "man eaters of Tsavo" had more acute food problems than did the lions of Kilgoris.

Fatal snake bites were surprisingly rare as well, and made me suspect that the danger of snakes in at least this part of Africa was also grossly exaggerated. Snakes were regular guests at our compound, but this may have been because of the abundance of poultry around the hospital. Even so, most bites we saw were those of harmless snakes, inflicted on terrified victims. Unfortunately on the rare occasions that victims of dangerous snakes were brought to the hospital, our antiserum had always expired.

The attack of the spitting cobra proved to be no myth, however. My new colleague Frank Froeling took me to a fifteen-year-old girl, whose face was etched by the venom of a cobra, that spat precisely in her eyes. She couldn't see initially, but to our relief her sight recovered and later even her skin. It is the beneficial effect of amylase, an enzyme in saliva, which makes the blindness temporary.

But perhaps the most curious snake injury of all was that contracted by a king-size ranger from Lolgorien, a nearby town, who jumped right off the ground at the sound of a hissing behind his calves. His Achilles tendon, torn by the abrupt movement, consequently had to be sutured.

So which are the truly dangerous animals of the African savanna? Well, beware the buffalo! I gradually started to perceive that buffalo, loved by poachers for their meat, were the most vicious animals around Kilgoris. The rangers described the behavior of males as unpredictable and sneaky, warning that they are most dangerous when you are unaware of their presence. An unseen beast wallowing in the mud behind a bush will launch into an unprovoked charge, perhaps in disgust at your nonchalance. They can cause serious injuries.

My first buffalo victim arrived on the minor operating table at eight o'clock one chilly morning. A shiny speck of blue was highlighted in the gored wound by the operating lamp.

"Do you see the kidney in this shitty mess," Jacobus grumbled. We had a tedious job cleaning out the dung from the musclerags. Although the buffalo had made a mess of this patient, after a few months, he recovered.

More nasty was the history of a fifteen-year-old Maasai girl, who had been attacked by a buffalo and came in semiconscious. A large wound overlay her intact, left collarbone. But more ominous was her distended abdomen indicating a ruptured spleen or something even worse. Her vital functions were okay, so we didn't lose time and opened her up. About a quart of sour milk gulped out of her abdomen. After some rinsing a huge hole in her stomach became visible. We repaired the stomach and cleaned her thoroughly inside. But for God's sake: Where did a hole the size of an apple come from? Besides the wound over the collarbone the girl showed no injury at all.

Only then did it occur to us that the two injuries were related. The buffalo must have flung the girl in the air, stabbing his vexed horn under the collarbone, down her ribs, and penetrating her stomach filled with the healthy Maasai staple food (milk). Very happy to have solved the problem and optimistic about her chances of survival, we went for dinner. The next morning she was dead. During the night her lungs filled with ooze caused by the lung contusion and she suffocated.

Bernard Grzimek in his famous book on the Serengeti questions the malign character of the buffalo: "A buffalo is as dangerous as a normal cow. He only attacks when wounded or annoyed," he writes, after his visit to a dying man (supposedly a poacher), who had been gored in his abdomen. He may have been biased by his love for nature. Buffalo-induced injuries ranked top of the list of the category "injuries inflicted by animals" in the hospital records year after year.

And they don't only attack people: "It was probably your ginger hair, or your freckles," I shouted to our Irish volunteers whose car had been gored by an angry buffalo, causing considerable damage. These blunt remarks (the result of Dutch straightforwardness, not always appreciated) were a little obnoxious, but I couldn't restrain myself. The hole in their car was impressive.

Once a most peculiar buffalo injury turned up on our operating table. A man had been gored on his way back from a wash in the river, and was brought into the hospital by his family. On examination we found a very short loop of large intestine bulging out of his abdomen, exactly midline above his umbilicus. The externalized loop had two small perforations. Already a considerable amount of feces had come out of the holes. Except for the feces it was clean and neat; I mean the skin was intact and precisely surrounded the bowel and virtually no blood had seeped around the injury. Forgive me my professional enthusiasm, but the buffalo had not only performed a standard colostomy; he did it at exactly the right spot in an all-time record of a split second. After closing the perforations in the colon, we continued with the exploration of the abdomen. It was virginal inside and the anatomy could be properly restored.

Thinking back on these incidents many years later, I decided to look in more detail at the hospital statistics. There were some interesting trends over time. For example, the increase in car accidents says much about rural development. In 1982 a tarmac road was finished leading straight from Kisii to Kilgoris, and the number of car accidents from that date rose significantly. The fact that Maasai suffer so much more greatly from speeding vehicles than enraged wildlife underscores the misconception Westerners have of the danger of wildlife in Africa. I also noted that the declining injuries inflicted by buffaloes and other animals can largely be attributed to their losing the battle against poachers. Indeed there were no rhino injuries in the eighties, because they had been poached from the area.

"Grievous harm" was the legal description for injuries inflicted by animals, and people could try to get some compensation by submitting a P3 form to the local court. The wildlife department would then pay the family of the victim. I never liked filling out P3 forms; indeed I never even asked how P1 or P2 forms contributed to the course of justice in Kenya. The consequences of filling out the P3 was that I would be asked to appear in court to comment on the medical report. This meant leaving the hospital with only one doctor, a disturbing outcome given the triviality of the court's compensation in comparison with the needs of the throngs of patients who, from dawn to dusk, waited quietly in our compound, or lay sometimes two to a bed, waiting to see a doctor.

And people desperately wanted to see a doctor. It was always rewarding to help, even though many of the afflictions we treated were preventable. Prevention, however, is not easy for the Maasai, given the strength of their traditions and the simplicity of their living conditions. The importance of prevention can only be conveyed through extensive education programs, and these are not yet very effective in the area. I therefore find myself more frustrated by a smoker in Holland who complains of his chronic bronchitis than by a Maasai who defaults on his tuberculosis treatment.

Giving medical care to victims of wildlife was my contribution to conservation. These people needed and deserved care, and often we could help them in what must have seemed miraculous ways. I admired and wanted to reward the tolerance of the Maasai for wild animals, given the constant threat of danger and the occasionally excruciating agony that wildlife posed. Just consider the reaction of Europeans or North Americans to the proposal of reintroducing wolves and bears into their neighborhoods.

Knowing that a majority of the Maasai still die in the bush and having seen day-in and day-out the pain they suffer in the hospital, I can only be impressed with how they manage to accept life as it is. Is it in respect for the course of events that they leave their deceased for the hyenas and jackals? I wouldn't mind such a funeral as a last contribution to wildlife, but then, I guess, I am a hopeless romantic.

Section III: Coping with Adversity

Paper Trail to the Rain Forest

by Lisa Halko and Marc Hauser

A RULE OF thumb exists for bureaucracies all around the world: the simpler the task, the more permits required to do it. Our task was to drive from Nairobi, Kenya, to western Uganda.

We arrived in Nairobi and passed through customs with a "*tafedhali*" [please] or two. The used Land Rover we'd just purchased was ready to be driven across Kenya and Uganda to the Kibale Forest where Marc was to study chimpanzees. But we wouldn't be allowed to bring it over the Ugandan border without an import permit from Ugandan customs, for which an export permit was needed from Kenyan customs. We couldn't get that until we had a Kenyan road permit, which could not be obtained without proof of ownership, which in turn counted for nothing unless we could show that it was recorded in the vehicle's office log book. The necessary log book was somewhere in the Ministry of Transportation.

"I have sent my messenger," said the dealer, "and he will return at three this afternoon. You come here then, and I shall give you the log book."

At three we returned. At four the messenger arrived.

"I am going to the Ministry now," he said. "You wait here and I will bring you the log book." Two hours later, the ministry was long closed and the messenger must have gone home for supper. We went home for ours.

The next afternoon we were back, and eventually so was the messenger—without the log book.

"I am going to the Ministry now," he said.

We went along, following him into an office where two young men in blue jackets were sitting at small desks covered with red folders. They were log books.

"Excuse us," we began, "we have just bought a vehicle. Would you tell us, please, how to get the log book for it?" Then our guide stepped in to take a firmer approach. As he harangued them in several languages, the two men slowly opened log books, stamped them with purple ink, and stacked them. They paid no attention to us at all. A very young man came in and took a stack of log books away. One of the occupants opened a newspaper, the other picked up a telephone and talked to a friend in Kikamba and Kiswahili. He made an appointment to meet the friend, chatted about some family business, hung up, and went back to stamping and stacking. After a while our guide stopped talking, or rather drifted into silence, occasionally grumbling, "*Saa hii!*" [Right now!]

Finally one of the men looked up.

"Good afternoon," he said.

"Good afternoon," said Lisa. "We have come to pick up our log book."

"I am very sorry," said the man. "It is much too late in the day to do anything. You see that it is almost four-thirty now. You should have come earlier."

"I see," said Lisa, "we'll come back tomorrow. Could you tell us please, if this is where to come for the log book? Or should we go to another office." We were having some trouble restraining our advocate, and hoped to give him the slip and start over without shouting.

"I am sorry, it is much too late for this," said the man. He went back to his newspaper.

The next morning at eight we were waiting outside the ministry when the doors opened. An official in the lobby told us that the log book could not possibly be found, and if found could not possibly be picked up. It would be mailed in six to eight weeks. To Michigan.

Michigan? Yes. The University of Michigan was administering the grants for our research, and our dealer had filled out the forms with the Ann Arbor address.

Six hours and eight offices later, we had progressed considerably. We had a yellow bit of paper which when signed by two unidentified officials, would allow us to ask a third person where our log book was. The person who had given us the paper sent us to the Assistant Under Registrar for his signature. In his outer office a clerk told us that no one could help us; they were all at lunch. Then the same clerk told us that before the Assistant Under Registrar would sign, we must get the signature of the Vice-Registrar in charge of export. In the Vice Registrar's office another woman told us vehemently that the Assistant Under Registrar must sign first. Back we went, to be met with annoyance by the initial clerk who ordered us not to return without the Vice Registrar's signature.

After several rounds of this, a visitor from another office, with her arms full of log books, showed us to a door. Behind it was a very old man who asked our names. In addition to our names, we told him the license number of the Land Rover and how much we'd paid for it. We told him our address, the names and addresses of the friends we were staying with, the numbers on our passports, how long we had been married, and why we didn't have children yet. We told him how long we had been waiting and what our purpose was in Uganda while he wrote it all down. Finally he said, "You wait here."

When he came back, he took us into a room the size of an airplane hangar which was filled with small desks. On each desk were three or four stacks of log books, any one, or none, of which might be our precious *sine qua non.* Leading us down a row of desks we felt like anxious lovers walking down the casualty ward after battle, looking for our sweethearts. He stopped at a desk, made an explanation in rapid Kikuyu, and left.

"Sit down here," we were told. We sat.

The man at this desk had a gold ballpoint pen which he frequently took out of his pocket. He clicked it as he read his paper, drank his tea, and looked us over. He got up and left.

It was nearly four o'clock and we were afraid that he would come back to declare the day over. But he came back with a piece of green paste-board in his hand. It was our log book.

We wanted to grab it and run out of the hall and down the seven flights of stairs, pushing anyone who got in the way. But we controlled ourselves, looking wary as mother cats do when you pick up a newborn kitten.

"Now," said Gold Pen, "what is the trouble here? When this log book is ready, it will be mailed to you. Why don't you wait for it at home? You don't need it. It is only for when you sell the vehicle."

"It is also for traveling," said Marc. "To cross a border, we must have the log book."

"Where do you want to go, now?"

"We must go to Uganda for a while."

"Now this is bad," said Gold Pen. "That is a very bad country. Why don't you just stay here?"

"But our job is to go there. If we do not go, we will not be paid. So please, *Mzee* [distinguished sir], may we have the log book?"

"Now, you must wait until this log book is ready. It is not ready. It will be ready in some weeks."

"What does it need? Because they told us it was ready."

"It has not been stamped."

"If you would tell us who needs to stamp it," said Lisa, "we could ask that person to stamp it today, so it will be ready."

"Aha." Click, click, click went the gold pen.

"Can *you* stamp the log book?"

"Now," said Gold Pen, "I cannot stamp this log book, because I do not have that stamp. Let me see whether I can get it. But first, I must have my tea."

And he went away again, taking our log book with him.

It was about three-quarters of an hour later when he returned.

"Now," he said, "let me see about that stamp."

He looked in the top drawer of his desk. No stamp. He went to a small office in the outer hall, and there was no stamp. Looking around, we noticed that unlike the people in all the other offices we'd seen, the people at these desks were not stamping log books. No one seemed to have a stamp. No one even had a purple ink pad.

Mr. Gold Pen sat at his desk. He clicked his pen. He folded his hands on the log book. He looked into our eyes. We looked back. It was a quarter to five. Without a stamp, it seemed, we would be starting fresh on Monday, never to find Gold Pen again. The log book would disappear again onto a pile of log books on someone's desk, and we would grow old and die in the stairways of the Ministry of Transportation.

"I will just sign it," he said. Click, click, click. Scribble, scribble. He put the log book into Lisa's hand and opened his newspaper.

A week later we were in Kampala getting more permits. In addition to the permit to bring the Land Rover into the country, and the permit to keep it in the country (renewable every two weeks—we'd have to drive eight hours each way to keep it current), and the permit to bring dollars into the country, and the permit to change money, and the permit to buy rationed "necessities" (kerosene, petrol, sugar, soap, beer), and the permit to enter, and the permit to work, and the permit to bring in scientific equipment, and the permit to carry an extra five gallons of petrol in a jerry can—in addition to all these—Marc needed the National Research Council to approve his program and the Forestry Department to grant him a forest permit. We thought we'd start at the National Research Council.

Everyone laughed at us after our experience in the Ministry of Transportation, saying we simply didn't know how to grease a palm properly. Bribe, bribe, bribe, they said, and in Uganda, bribery is even more necessary. Steeling ourselves against the conviction that we'd be offending countless law-abiding bureaucrats, *magendo* (bribery) would be our creed.

We climbed the steps of the National Research Council and walked into the first office. There was a sign on the wall, right under the picture of President Museveni. "No magendo here," it said, "we do not put money before God."

"Good afternoon, madam," said Marc, "I applied for a research permit six months ago. They said I would be able to pick it up here."

"Good afternoon, sir," said the woman in the office. "Next door down."

"Good afternoon, sir," etc. said Marc again.

"I am so sorry!" said the man in this office, "the officer who issues these permits is not in. You may come back next week."

"I believe my permit has been issued," said Marc. "I was told that it would be ready."

"Ah yes. You see, the Council has changed its members. Now they require that you have a permit from the Forestry Department first. Then, once you have permission to be in the forest, we can decide whether to approve your project."

So Marc reapplied for the permit, which required a new set of six pictures and a three-page form filled out in quadruplicate. The following day we went to the Forestry Department which was located in Entebbe, a forty-five minute drive from Kampala. Although the top official at the Forestry Department was not expected to arrive until ten in the morning, we left at six so that we would be waiting when he arrived.

"Good morning, madam. We would like to see someone about obtaining our E-permits for working in the Kibale Forest in western Uganda."

"Good morning to you! The Chief Forestry Warden is the one you must see. He will be in at 1:00 this afternoon. You may wait here."

So we did. It seemed that the working day in most offices was from 1:00 P.M., when lunch ended, to 3:00 P.M., when people began to wind down in preparation for going home at 4:00 P.M. We found out later that no one had been paid their government salary for six months to a year, and even payment in full wouldn't support one person, let alone a family. So in order to eat, everyone had a farm to tend. Uganda was then a nation of subsistence farmers who taught, delivered mail, repaired roads, bore arms, judged lawsuits, investigated crimes, administered farming co-operatives, and generally ran the country in their spare time, for free. Talk about public service!

Luckily the wait was interesting. Though the building was as shabby and dark as most other government buildings, its walls were covered with old maps of all the major forests in Uganda. Most of the maps dated to the 1960s and charted non-existent forests that had already disappeared from heavy encroachment and logging.

At 1:30 we saw the Chief Warden walk past us into his office. He was a tall man, over six feet, who must have weighed close to 250 pounds. The secretary motioned us politely towards the Warden's door.

"Good afternoon, sir, I am Dr. Hauser and this is my wife, Lisa."

"Dr. Hauser, aha," said the Warden. "Why have you come here to Entebbe?"

"I have come to study chimpanzees in the Kibale Forest and understand that we need E-permits," said Marc. "Last summer I came here to apply for them, and brought a letter of reference from the Zoology Department at Makerere University. Is it possible to pick up the permit?"

"Dr. Hauser, aha. I don't remember your proposal. Really, you scientists must learn to keep us informed. Let me check my files."

"Yes, sir," said Marc. "I submitted a description of the project last summer, and sent another when I wrote in the spring. Perhaps you would like this copy here." Marc put it on the desk.

The Warden continued thumbing slowly through an ominously tall stack of papers. After reaching the bottom he said, "I don't see your file here but I remember talking with you last summer. Where can these papers go? Now, let me see your permit from the National Research Council. Of course we cannot allow you to work in Kibale until we know that the NRC has cleared your project."

"Well you see," said Marc, "the NRC has not yet granted the permit because the officials there say that we must have the E-permit from the Forestry first. They say that the NRC has changed its members and this is a new policy."

"This is a policy that I know nothing of. I am afraid that you will have to come back some other time when I have discussed things with someone at the NRC." The Warden stood up, shook our hands, and said good day.

"Of course *you* will not need any permit," he said helpfully to Lisa. "You are just his wife."

We asked when we could return. The Warden said that the NRC met only four times a year, so he recommended waiting at least a week. Maybe a month.

A month in Kampala doing nothing! We were tempted, since we clearly would never reach Kibale or see a chimpanzee, to abandon the project. Instead we had dinner, which cheered us up considerably because it was *tilapia* and we wouldn't be having fresh fish in the forest.

The next day we went back to the NRC to find out about the order in which the permits needed to be obtained. This time a new clerk was behind the desk.

"Good day to you!" said the man behind the desk.

"Good day to you," we said. "We would like to find out about permits to work in the Kibale Forest on chimpanzees."

"You are Dr. Hauser, am I correct?"

"Yes, that is correct."

"We have your permits right here. In fact, it has been ready for the past six months. I understand the man you spoke to yesterday said that you would have to reapply. Sometimes these new people can be very incompetent. Here is your permit. You may now go to the Forestry Department and they will issue your E-permit."

We shook hands gratefully and drove straight to Entebbe, where the E-permit materialized just as miraculously. We visited our counterpart from Makerere University to tell him that we could give him a ride to Kibale the next day, rather than waiting for the indefinite future. In the morning we crowded into the Land Rover's little front seat.

It was rumored in Kenya that the Ugandan road blocks were staffed with fierce soldiers who would let us by only if their threatening demands for cigarettes were satisfied. We bought a whole carton, though neither of us smoked, and hid it under the seat. At the first road block just inside the border, Lisa put a few cigarettes in her shirt pocket. We drove on, were waved down, and stopped. Five soldiers sat in the dust under a tree, wearing tattered civilian clothes. Each was distinguished as a soldier by a single article of green clothing—one with ragged shirt, another in a pair of boots, and another with a cap. A boy of about fifteen approached, handling his machine gun too recklessly for our taste.

"*Habari gani?*" [How do you do?] he asked.

"*Mzuri sana, na wewe?*" [Very well thanks, and you?] said Marc. He was speaking to the words rather than the tone, which was as peremptory as any fifteen-year-old could make it, even without a machine gun. We prepared to get out of the vehicle, unpack everything, hand over our cigarettes, and escape with minor robbery.

"Do you have anything to read?" asked the boy.

We were touched. All the soldiers at this roadblock, it turned out, were schoolchildren or young teachers. They had interrupted their studies to join Museveni's army. Having won a guerrilla war and taken the capital, they were left to sit in the hot sun with nothing to do. They didn't want to smoke or steal; they wanted something to read and discuss. We unpacked our things, no longer fearful, but in hope of digging up something with news in it. We offered *Newsweek* and *Time* magazines and some paperback Shakespeare plays.

Leaving Kampala for Kibale, we were better prepared. After purchasing some newspapers, magazines, and several volumes of Museveni's essays, we set off bouncing our Land Rover over the rutted dirt road, stopping at nine military road blocks along the way. Though we traveled less than 180 miles, the trip took over eight hours.

Upon our arrival in Kibale late in the afternoon, we were greeted by Gary Tabor, acting director of the Kibale Forest Project. Gary showed us our new house, a tin-roofed, brick structure in a forest clearing. We had brought special food—chocolate and noodles—from Kampala, and Gary ate with us that night. We immediately felt intimate with him which was partly because he was a remarkably forthcoming, straightforward person, and partly because Lisa stepped in a swarm of *siafu* (a species of African army ant) while cooking in the dark, and had to tear off her clothing in front of him just three hours after having met.

During the following eight months Marc learned about chimpanzees and worked with Richard Wrangham, a primatologist, to establish the meaning of the chimpanzees' vocal repertoire. The chimpanzees, as it turned out, were too elusive to gather much data. But Richard never came back from a day of observation without an exciting description of some species.

Marc slogged through the steep, wet forest looking for chimps who made up for frequent absences by their amazing on-site behavior. Marc got to see a young male adult make a bridge of his body so that his old mother could walk from one tree-top to another. He witnessed a five-day orgy when six females came into estrous at the same time and situated themselves all in the same huge, fig-laden tree. Remarkably, there were times when the males were much too interested in food to be interested in the females' solicitations for sex.

Lisa worked with the public health team from a local hospital, going to remote villages to immunize children. Poor kids! Many had never seen a white person before and were fearful because their mothers had told them that "*mzungu* (white person) will eat you if you don't behave." It was a great publicity stunt for the clinics, as the children brought their mothers and siblings to see the woman with the yellow hair weighing babies on a butcher's scale in the marketplace. But some were so frightened of the strange creature, especially when it began to stick them with needles (BCG in one arm, measles in the other, DPT in the thigh), that only mental recitations of infant mortality statistics prevented Lisa from taking pity on them and leaving them unimmunized. Oddly, no one ever demanded to see a nursing license; willingness to pitch in was the only relevant qualification. Except for trips to Kampala to keep our permits current, the only time we needed our papers was to buy extra petrol which was only necessary because we had to drive to Kampala to keep our permits current.

The forest was beautiful, the people were kind, and nothing had anything to do with permits. It was with regret that we drove back to Kampala where Marc was to begin teaching a two-week course in Animal Behavior at Makerere University. After the course, we would drive back to Kenya and then fly to New York.

Two days before our drive to Kenya, we went to the Nakasero Market where non-Ugandans shopped for costly imported produce. We wanted to make a feast for the expatriate friends who had been kind to us. We had already roasted a goat for our neighbors in Kibale and were looking for green beans and peas (not exactly tropical vegetables). Unfortunately, our very presence at the overpriced Nakasero Market loudly proclaimed that we were carrying enough money to shop there. We were duly pick-pocketed. We discovered this later when looking for our plane tickets while trying to reserve a special seat for Lisa who was eight months pregnant and took up too much room.

The tickets had been in a little black purse in Marc's satchel along with his passport and wallet. All the permits were there, too. And of course it was the little black purse that was missing.

Well, you don't replace paperwork without a lot more paperwork. At the American Embassy they told us that we could get a passport in a few hours, but first a police report needed to be filed. It was already four in the afternoon—much too late to tackle the police station. The next day was Friday. Mark had to lecture all morning, and it was the business day before our plane left from Nairobi. Many permits needed to be reissued. How would we ever get the papers necessary to leave the country?

At 8:00 A.M. Lisa went to the police station to report the theft and get a copy of the report. Of course, the official policy is that the reports are not copied. But a kind officer agreed to write a letter explaining why we had no papers to show to the border guards and the American Embassy.

Marc lectured, then spent a moment telling a colleague, Gilbert Isabirye Basuta, what had happened. As Isabirye commiserated and asked for details, another man approached. Isabirye introduced him as Christopher, a technician. Christopher had been listening to Radio Uganda.

"Last night on the radio," he said, "I heard the announcer say that a purse was found with log books to a Land Rover as well as a number of other papers. Could it be yours? It is with the chairman of one of the resistance committees. I don't remember which one."

The choice: spend the day tracking down the announcement and the purse, which might or might not be ours, or spend the day getting the police letter and the replacement papers, which might or might not be possible. Christopher said he would help. He and Marc picked up Lisa from the police station and headed for Radio Uganda.

Upon arriving at the receptionist's desk we asked if we could look at the messages from the previous day and perhaps speak to the announcer who was on the air last night. The receptionist told us that the announcer was sick at home but that we could look through the messages. We were led down a long corridor and placed in front of a massive heap of announcements where we each took a stack and began thumbing through notes written in thirteen of Uganda's twenty-seven languages. But, none of the messages had anything to do with our purse. The receptionist told us that there was another stack of messages locked away in a cabinet. However, the announcer from last night was the only one who had the key.

Christopher asked where the announcer lived and the receptionist reluctantly told us. He knew the place and we drove to the outskirts of Kampala.

Kisa, the announcer, lived in a shack behind a school. He answered our knock, "Yes, may I help you? Please excuse me, I am quite sick."

Christopher explained, "We need your help. We understand that you are the announcer for Radio Uganda who read the announcements last night.

Kisa answered, "How do you know that?"

"Do you remember reading about a purse found with several permits including a log book for a car?" asked Marc anxiously.

"Yes, I do seem to remember reading something about a purse but I read over fifty messages last night."

"Would you be willing to return to the station with us to open the cabinet where last night's messages are?"

"I suppose I could do that," said Kisa, coughing a bit. "But how do you know who I am?"

"Oh, you are a celebrity . . . the voice of Uganda," said Christopher.

Back at Radio Uganda, Kisa opened the cabinet and there, close to the bottom of the stack, Lisa discovered our message, or so it seemed. Christopher, who was beginning to appear omniscient, knew where the district was and where the chairman lived. After dropping Kisa off at his house we drove to the center of town and parked. Christopher led us through the back ways of Kampala, past small markets, government buildings, old tin houses, and outdoor butcher shops with meat hanging from hooks and, in some cases, carpeted with flies. It was mid-day, and the alleys were filled with cook-pots of soup, braziers of meat, and griddles heating flat bread. There was barely room to walk, because every space not filled with a cooking utensil was filled with a person eating.

We found the chairman in a decrepit shop, talking with a few women who were selling smoked fish. Christopher asked the chairman if he had recovered a black purse.

"Well yes," said the chairman, "but actually I only reported it to the radio. This woman was the one who found it but she did not know what to do. You see, she saw two boys take the purse out of your bag, remove all of the money and then throw the purse on the ground." The chairman then reached into a filling cabinet and pulled out our black purse. "You see," said the chairman, "I think you will find all of your papers here."

We couldn't believe it. The only items missing were 3,000 Ugandan shillings (about ten dollars) and our checkbook. All of the permits and Marc's passport were safe in our hands.

The ever-resourceful Christopher knew a good restaurant, and we treated him to lunch. We were all as cheerful and tired and triumphant as anyone could be, and more grateful to Christopher than we could express. That night we slept well and woke up early so that we could reach Nairobi before dark. Richard Wrangham's wife, Elizabeth, and their two sons were leaving Uganda as well. Their nanny, Jennifer, was traveling with us to Nairobi to meet Elizabeth and the boys when they flew in from Kampala.

At the first road-block the soldier asked us whether we had "any bad things" in the vehicle.

"Oh, no," we all said, and he waved us on. We arrived at the Uganda/Kenya border at noon and drove up to the little booth. We had our papers in order. We each had a passport, and handed those over as we said hello. We were well-prepared to get out of the car and wait in several lines at several different stations, to have our money counted and compared to numbers on the foreign-exchange permits, and to be questioned about the length of our stay and the time of our return.

We never even got out of the vehicle. The guard saw the Makerere sticker which was pasted on the side of the car, and motioned for us to stay inside. He opened Marc's passport, glanced at it, and stamped it along with Lisa's and Jennifer's.

He said, "You are Dr. Hauser from Makerere University, and these are your two women. You may cross."

Well, there you are. With all of the papers or none of the papers, an academic polygynist goes where he likes.

My Family, Food, and Fieldwork

by A. Magdalena Hurtado

FIELDWORK AMONG PEOPLE whose behavior and ideas about the world differ greatly from our own cannot only be a very funny experience, but also a revealing one. We enter new social and belief systems with misplaced and unrealistic behavioral expectations of ourselves, and of the people we hope to befriend. Frequently, we merely watch, and learn a great deal about people's social preferences, concerns, ideas, feelings, and world views.

I have spent thirty-three months in the field. I did most of my earlier fieldwork with a friend (Hillard Kaplan) and my husband (Kim Hill) and later, with our daughter (Karina) as well. We've worked with lowland South America indigenous peoples, mainly the Ache of Eastern Paraguay and the Hiwi of Southwestern Venezuela, who depend on foraging for their sustenance to varying extents, and for whom food is an obsession. Food became my obsession as well, and learning how to satisfy my cravings ironically taught me a great deal about how married women manage to procure meat from men in these societies.

In addition to food, social relationships are also an obsession for the Ache and Hiwi, as is probably the case in all societies. I found that becoming socially competent under field conditions is almost impossible due to the demands of data collection and the numerous other self-sustenance tasks one needs to accomplish. However, I made every attempt to attain at least a level of social competency that could be tolerated by the people we lived with.

The Ache are a lot of fun. They love to laugh, tell jokes, tickle one another, and show physical affection in many ways. It is not at all uncommon to find Ache sitting very close to one another, spontaneously cleaning insects off each other's bodies, and feeling quite at ease with physical contact. Up to that point in my life, I was under the misconception that the Venezuelan culture I had grown up in had reduced individual's immediate spatial boundaries to a minimum. The Ache eliminated that last bit of space.

The Ache were traditionally hunter-gatherers. We worked with a band that had settled over the course of the 1970s in an agricultural settlement, and depended on corn, sweet manioc, and store-bought goods for much of their subsistence. However, they frequently left the settlement to hunt and gather over weeks and months at a time in the surrounding forests.

The events I am about to describe took place during my first visit to the Ache in 1981. We lived with this band over a period of nine months in foraging camps comprised of three to seven families. We stayed in these camps for weeks and months at a time, and spent very few days at the agricultural settlement. These were my most interesting experiences because I spent so much time in the forest.

Because Kim had lived several years with the Ache prior to my first visit, it was relatively easy becoming "socially accepted." Being associated with Kim was certainly not enough, however. The Ache watched me very closely to see if I learned to do things the Ache way, and when my mastery of skills was flawed they ridiculed me with delight. It took a while, but I finally learned to laugh at my awkwardness since no matter how hard I tried, I simply couldn't do such simple tasks as gathering firewood, starting a fire, keeping a fire going, and cooking efficiently.

On my first foraging trip, I had a very vague notion of what to expect. I knew that we would probably wake up before dawn every day, start walking to our new camp not much later, and arrive sometime between two and six in the afternoon. I was worried about stepping on snakes, falling off makeshift bridges, or losing my data notebook or pens. I wondered if I could survive an unpredictable number of days eating monkeys, peccaries, armadillos, palm larvae, palm starch, and other foods I had never tried before.

On that first day I made friends with an older Ache woman. She stayed with me and this made me feel secure. I opportunistically chose her as my focal person (the individual whose behavior I would sample over that entire day). By using her, I violated an important rule in data collection: sampling individuals at random. It turned out that this violation of protocol was insignificant compared to all the other biases I introduced those first days. I collected less data when mosquitoes and gnats turned writing into torture, and collected still less, if any, data when we stood under the hot sun and sweaty hands made my Bic pens useless. Eventually I learned to avoid these problems by wearing several layers of clothing when the bugs were bad, by keeping my data notebook in a small plastic bag, and by using a handkerchief frequently to dry my hands. The initial data were so poor that we had to exclude them from all our analyses.

My memories of foraging trips are imbued with hunger. I always felt as if I didn't have enough food; and I could never tell when I would get it. Between "bouts" of data collection I would oftentimes think about the two pieces of candy and handful of nuts I had hidden in my knapsack for an after-dark meal.

By the second day, I had eaten most of the candy I had brought and tried to convince Kim and Hilly to give me theirs without success (we each kept a few hidden in our daypacks). Consequently, that evening I was so hungry that I managed to eat monkey meat even though the night before I wouldn't take a single bite.

That night I learned that when a woman is accompanied by her husband, she is not given meat directly, but is expected to share with her spouse. Kim had spent the day with a hunter who decided to return to camp at dusk. There were a few who had returned earlier in the afternoon, meat had been cooked and several people were eating. I had to wait until dark to eat, and not until after Kim had bathed. Up to that point I had been very happy about the benefits of having a spouse in the field, not realizing that when it came to meat, husbands could be a liability for women; the few, single women in our foraging trips were shared with directly and plentifully, but not me. As I quickly ate a piece of armadillo I had hoped to eat hours ago, I wondered about the generalization one often hears in introductory anthropology courses: that food is equally shared in hunting and gathering societies.

I was much more preoccupied with food than were Hilly or Kim. Interestingly, the same dynamics could be found in the Aches. In fact, Ache women were far more obsessed with food, especially meat, than the men seemed to be. At night I packed away the meat that Kim and Hilly had left over so I could eat it during the day whenever I was alone. I had to time it just right; if I waited too long maggots would grow all over the meat. I soon discovered that the Ache women and I were not much different after all. They also hid pieces of meat in their baskets, and clandestinely consumed it throughout the day.

Several months later I had another interesting experience with food that made me question, for the second time, the widely-accepted egalitarian characterization of hunter-gatherer societies. It was one of my favorite times of the year: *"kurilla"* fruit were abundant, ripe, and sweet. We spent several days at a beautiful grove, with towering trees, clean forest, and a flat, sleeping area. The first day at the grove I went fruit gathering with several women and picked as much fruit as possible because I was so hungry. When I returned to the camp I started eating as much and as fast as I could in order to go out again and get more for Kim and Hilly. Unfortunately, the feast lasted only a few minutes: An Ache woman yelled across camp, with others joining in, that I stop eating the fruit. They said I had to save it for Kim and could only eat more after he had eaten. While ceasing to eat, I worked to convince myself of the merits of cultural relativism—the belief that the values of all cultures are equally good—to deal with this confrontation, and actually managed to stop eating. Fortunately, Kim was never as food stressed as I was, so I ended up eating a lot of the fruit after dark. I didn't want to get scolded again.

Besides being concerned with food, I spent a lot of time trying to figure out ways of communicating with the Ache women. I hoped to develop some friendships, but I found that the women weren't very talkative on foraging trips. Walking from one camp to another with huge baskets, children, and pets on their backs, takes a lot out of women. When we would stop to rest, they would lie down or sit alone, sometimes brushing the insects off themselves or their children, eating, or grooming other women. When women forage, they rarely coordinate complementary activities with one another, as opposed to the Ache men whose hunting is primarily cooperative. The women seemed to be more social in the evenings when men and women engaged in lively conversation, told jokes, and sang songs. As mentioned earlier, they also enjoyed themselves a great deal when I would unintentionally do very silly things. On my first foraging trip, I sat on the same thorny palm tree trunk twice. The women laughed so hard at this, that tears formed in their eyes. Such events made me feel embarrassed and frustrated. But as I learned more about the Ache, I realized that they laugh at each other in the same way.

Peccary, coati, and paca hunts were also a great occasion for the women. I only saw women participate in these hunts on four occasions. Women would divide the labor such that one or two would stay behind with the children while the others would help the men locate prey in the thick forest or in burrows. During these hunts I learned that Ache women know a great deal about hunting, and yet they rarely do it.

I saw Ache women come together in yet another and unexpected way. In one of our early trips, when my communication skills were horrendous, I spent the day with one of the more influential Ache women, Vachugi. She had a forceful personality, participated in group discussions and, not surprisingly perhaps, was also physically very strong. We were walking through an area with plants full of small thorns, it was very hot, and there were many mosquitoes. The women stopped to see a dead peccary that the men had just killed. Vachugi sat by the peccary and started singing and crying very loudly. The rest of the women joined in. I couldn't understand what was going on, of course. They held each other and cried while I stood by. Then suddenly, they stopped the weeping completely, as if though nothing had happened, and started to tease one another. Later on I learned that they were crying over a deceased relative whose name had been peccary: *Chachugi.* Interestingly, the same women did not display their grief overtly when a baby died at the settlement: only the mother could do so. Once again, I got scolded this time, not over food, but for crying.

I also tried to learn how to become friends with Ache men. This was much trickier because I didn't know what incorrectly-timed, subtle behaviors could be unintended sexual invitations. Among the Ache, there are no spatial boundaries between the sexes. There is considerable flirting and touching among men and women who are not married. Men playfully grab women in public, pretending to pull them into the bushes. Women have considerable leverage in these public flirtations, both stopping and initiating them whenever they want. I was never quite sure, however, what behaviors were clearly flirtatious, and what others could be interpreted as a serious pursuit. So despite my caution, and very unintentionally, I ended up doing the latter.

After spending seven months with the Ache, I started feeling stronger and more comfortable on our foraging trips. I had learned to get firewood with some level of competency, to time data collection, cooking, washing, etc. in a more manageable way, and to communicate better. So I began to take more risks in my interactions with the Ache. I had learned enough key phrases, I thought, to get out of almost any embarrassing situation. On one of these later foraging trips, I decided to try carrying a four-year-old boy, whose father had recently been left by his wife, on my shoulders. I felt sorry for Javagi because most Ache children are carried by their mothers at that age, and he cried constantly. After carrying Javagi for a while, Bepuragi, a woman of high status, married to one of the best hunters, told me to make the child walk, and said several other things I didn't understand. I could only tell by the tone of her voice that she was very upset with what I had done. I forgot about the incident until a few days later.

After returning to the settlement for our usual period of rest and recovery, the boy's father asked me (in public) to move in with him. He was quite serious, and he also didn't seem to mind that Kim was standing only a few feet away from us. Perhaps carrying a man's son is a form of sexual invitation among the Ache, and Bepuragi had been warning me of this. I don't think I'll ever know.

Two months later we left the Ache. I had a baby, and returned to the Ache for a second visit when she was just a year old. Even though we had nice tents, a generator, reasonable food, etc., I was unable to carry out the research project I designed for this subsequent visit. Preoccupied with the cares of motherhood, I had to boil water constantly for her formula, and continually wash and boil bottles and other eating utensils. In spite of this, Karina had stomach problems constantly. The humidity kept her awake at night, so that I could never sleep more than one hour at a time. During the day, I had to carry her often in order to keep her from wandering into the forest. She was very attracted to the forest because it was full of sticks that she would pick up and use to dig holes, or hit insects, or anything else that moved. Unfortunately, the Paraguayan forest is full of stinging insects and nettles, and like Ache mothers, I had to restrict Karina's activities as much as possible. Not surprisingly, I was ill most of the time during this visit. After accomplishing very little, I returned to the United States to write my dissertation.

After I finished my dissertation, Kim and I decided to start working in Venezuela since we could count on relatives to help us with childcare and other logistical support.

There, we learned from Roberto Lizarralde, a Venezuelan anthropologist, about a group in Southwestern Venezuela that depended primarily on hunting and gathering for its subsistence. Unlike the Ache, the Hiwi are savannah, riverine foragers. They live in the flat, extensive *llanos* of Venezuela and Colombia and take refuge from the merciless sun in settlements protected by the canopies of gallery forests found along the banks of rivers and streams.

The Hiwi economic system appears to fluctuate between brief intensive periods of horticulture and longer periods of foraging. When we first started working with the Hiwi they were full-time hunter-gatherers. However, by 1990, they had planted large manioc and corn fields and consumed few gathered foods.

The first time Kim and I visited the Hiwi we were brought to the settlement by government functionaries working in the local indigenous affairs office. Our journey began in a tractor with several Hiwi who were returning to their settlement after a visit to a small Venezuelan town. The ride took half a day. On the way we befriended a Hiwi man who invited us to stay with his family. We traveled by boat the rest of the afternoon and finally reached our destination: a small village of fifteen families located on a high river bank, amidst the gallery forest. Our new friend made room in his house for our hammocks and supplies, and within a few minutes we had moved in. We were offered wild roots and turtle soup for dinner, slept well, and then spent several days getting to know this Hiwi community.

The Hiwi are strikingly different from the Ache in many ways. There is very little physical contact in Hiwi society. Unmarried men and women seem to purposely keep themselves as far apart in space as possible. Girls play only with girls, and boys with boys. Adult men and women form separate groups such that the women sit together and the men sit elsewhere. Mixed sex groups are confined to husband-wife teams: even adolescents keep themselves segregated along sex lines. But most interestingly, men and women do not face each other when they communicate, and rather than speak, they yell to one another in opposite directions.

I learned this about the Hiwi after spending just a few hours with them. It was already too late: I had broken all the rules concerning male-female interactions. As an outsider, I got away with it. Fortunately, neither Hiwi men nor women seemed to be bothered by the way I communicated with the men, so after spending two days making friends we tried collecting some census and demographic data. I conducted interviews using a bilingual informant. We chose a bench under the shade of a mango tree in the middle of the village. I sat at one end, and my informant, a Hiwi man, in a very "un-Ache" way, sat at the opposite end. Then the interviewee, an older woman, sat on a tree stump with her back to my informant and her face down, about six feet from us. When Bawai started asking questions, he would yell them out and the woman would shout back the answer. Needless to say, we deleted all confidential and sensitive questions from our first "public" interviews. Later we built a small house where the informants could have privacy.

I never found myself in very embarrassing situations due to the physical separation of the sexes. I spent much of the time sitting with the women, foraging with them, processing food, and caring for the children. By the time we started working with the Hiwi, Karina had become a very important member of our field team. She broke all the rules, and the Hiwi loved it, except for one time when Karina clearly overstepped the boundaries between the sexes. Karina was four years old and she insisted that she wanted to have a bow and arrow to shoot small lizards with the boys. So she asked to borrow these tools and began practicing by shooting arrows at small objects. A few moments later, the defacto chief of the village came to have a talk with me (unlike the Ache, the Hiwi never yelled at me!). He informed me that if Karina played with the bow and arrow she would be sterile for life. Yesu was very polite, he explained that he did not want to tell us what to do but that the Hiwi didn't want to be responsible for this horrible health consequence. Karina was quickly convinced. After having played with many Hiwi infants, she had become very interested in her own reproduction prospects.

I never felt food stress with the Hiwi, and I think that this was partially because the Hiwi are central place foragers. They have a large settlement where they spend most of their time resting, eating, caring for children, and processing raw materials. Most foraging is done within a few miles of the village. This meant that we could keep store-bought food in our tents for several months at a time. But perhaps most importantly we did not need to accompany the Hiwi on their foraging expeditions to collect data: we only needed to clock them in and out of the camp and weigh all the food. The Hiwi consume very little of the food that they acquire in the bush.

Unlike the Ache, the Hiwi have a rich and interesting supernatural world. They felt strongly about protecting us from the evils of the savannah after knowing us for some time. When we first arrived, however, they were not quite sure if we had come from the good or the evil side of their supernatural world. One event convinced them of the latter.

In that first boat ride to our friend's village, we opened a can of small sausages and shared them with everyone in the boat. There were several Hiwi from an enemy village who refused to eat the sausages. Later we found out that they thought we were distributing children's fingers: we were cannibals. Throughout our first long stay in Bawai's village, the children refused to get close to us. We later learned that they thought Kim was the devil: *cauri*, a tall blonde blue-eyed devil. Fortunately, there were enough Hiwi men and women who didn't think we were cannibals, and we can only guess that they convinced the other adults. It took much longer to convince the children that Kim was not cauri.

Our frequent visits helped to convince the Hiwi children that we had nothing to do with cauri. I think that this fear was greatly reduced after my mother, brother, and I visited the Hiwi in April of 1986. My mother, Ines, joined me to collect data on parasite load and allergic reactivity as part of a long-term study of Venezuelan populations. We convinced my brother, Pablo, to be our male companion since it is generally unwise for women to travel alone in remote, rural areas of Venezuela. Pablo, a professional artist, took care of the photography for both projects as well. I was astonished to see how willingly the Hiwi, including the children, cooperated with my mother's request for blood and fecal samples and my brother's intrusion with his cameras. Everyone was extremely kind to us, but especially towards my mother. It was as though, simply by virtue of her age, she elicited a level of respect that Kim and I, and later my brother, never received. The Hiwi danced for her, and the de facto chief gave a speech in Hiwi and Spanish thanking her for her visit. With the

Hiwi, at least, it is an asset to take one's older relatives into the field.

Younger relatives were also an asset. In fact Karina was a star with the Hiwi. At our departure people showed the greatest concern over Karina leaving. We were told in no uncertain terms that we had to bring her back. Karina's intentions and interests were never questioned, and she was welcome in all the Hiwi households. She had the privilege of doing things with the Hiwi that we could not because of our age and status in the village. The women loved her fascination with babies. When she was only five years old, the Hiwi women would let her care for their infants. Unlike the Ache, Hiwi girls become skillful caretakers at that age! A real baby was far better than any doll Karina had ever had. Everyone also seemed to tolerate the way she broke most cultural rules: she sometimes sat among groups of men snorting the hallucinogenic powder *yopo,* and would ask to stick her finger in the powder to see what it felt like. Hiwi children usually stayed away from such gatherings. She would touch the sacred Hiwi *maraca* even though only a few of the men in the village had the privilege to do so. Karina also spent a lot of time getting her face painted, loved to participate in Hiwi dances, and to bathe in the river with the children. Finally, unlike us, she would sometimes join a Hiwi family to go to the bathroom in the savannah. Perhaps because of the evils that lurk in the savannah, no one goes to the bathroom unless accompanied by at least one family member, and usually the entire nuclear family went on these trips. At first, we suspected that the Hiwi had food hidden in the bush that they consumed on these outings, but Karina confirmed for us that these were not the picnics we had suspected.

Having a child in the field has been as advantageous as it has been inconvenient at times. And being accompanied on my fieldwork by my child, husband, mother, and brother has given me insight into the world views of the Ache and Hiwi, and eased my attempts at becoming socially competent enough to communicate with them.

Little Criminals

by Truman P. Young

REELING FROM THE sudden blow, Elizabeth looked at me with disbelief. As her milky under-eyelids slowly blinked up, my inklings of doubt were replaced with light-hearted satisfaction. I am not a violent man. But long-term isolation can make even a plant ecologist just a shade sadistic. After a long series of humiliating defeats, I was making a comeback.

It was 1981 and for the last four years I had been living in a tent at 14,000 feet on Mount Kenya, trying to decipher the secrets of tropical alpine plants. The locale was spectacular. Mount Kenya is an ancient volcano that was born when *Homo erectus* roamed the plains. At 23,000 feet, it was then one of the tallest mountains in the world. Its bulk was a great barrier in the path of the seasonal monsoons, and soon its slopes were clothed with thick wet forest above the arid plains. Above the limit of forest growth, snow fell profusely on the equatorial slopes. Great glaciers were formed, and these carved deep radiating valleys. Over the millennia, the upper 6,000 feet of the cone were ground away, revealing an impervious lava plug that, at 17,000 feet, now stands like a fortress more than 3,000 feet above the valley floors.

My camp was nestled on the wall of the Teleki Valley, literally in the shadow of the central peaks, Batian and Nelion. Every morning they hid the sun until mid-morning, and most evenings were bathed in a glorious rosy alpenglow. My tent opened onto this scene, and I never tired of it in all the years that I lived there. My open-air latrine had an even better view than the mammalogy outhouse at the Rocky Mountain Biological Laboratory.

The ideal research camp? Not quite. The end of the passable road was 4,000 feet lower, if it hadn't rained lately. If it had rained, the end of the passable road was anywhere from one to fifteen miles further down the mountain. It seemed that I always had the "luck" of a dry road coming up, and usually managed to get the vehicle to the very limit of the steep and winding road. It would then proceed to rain on the road during my entire field session. I lived above most of the weather, and so was oblivious to all of this. Every week or two, I had to go down for supplies (and a hot bath). By this time the road was often soaking wet, slippery, and deeply rutted. But I had no choice; I had to get the vehicle down. My winter mountain driving experience in Colorado notwithstanding, I have always disliked driving down cliff-hanging roads when a car's direction of travel is determined more by inertia than by steering. Such was the road to and from my study site. Of course, there was also a little hike.

The seven-mile trek to camp starts through a forest in which hikers were often treed by buffaloes and elephants. After climbing out of the forest, the "Vertical Bog" provides diversion for an hour or two. This massive quagmire is not actually vertical; but don't tell that to the exhausted hikers struggling half-submerged in mud in scenes reminiscent of the La Brea Tar Pits atilt. If you keep moving, you can reach the first potable water in four to five hours, still an hour or two below the campsites. By that time you are cold, wet, tired, thirsty, and the spectacular scenery is viewed through a haze of mountain sickness. At least that's how I remember my first ascent. With time and conditioning, the trip became more of a three-hour bore than a six-hour trial, but there was often a race with the afternoon showers to enliven the trip.

Tropical alpine climate has been described by Olov Hedberg as "summer every day and winter every night." Dr. Hedberg is Swedish, and that may explain his definition of "summer." Granted, if the afternoon sky was cloudless and the air was dead calm (an occasional combination), I could work comfortably in the intense high-altitude sun in a T-shirt and shorts. But a little shade or a light breeze reminded me of how cold the air itself was. On a partly cloudy day, I spent my time alternately disrobing and bundling up.

When I was not doing this quick-change act, I was studying the giant lobelias of Mount Kenya. Very different from their garden relatives, these giant lobelias resemble artichokes the size of basketballs. After growing from a seed the size of a pinhead for several decades, each one produces a ten-foot-high stalk with 5,000 flowers and a couple of million seeds. Then it dies. My task in life was to discover why.

Nature filmmakers are not generally attracted to this sort of thing. I am not complaining, but it does get old seeing someone you know every time you turn on the television, and it is always someone studying a big furry animal.

When you study large mammals, you have plenty of company. There are often other scientists around studying the same system, visitors (and filmmakers) are interested in your work, and you get to know the animals themselves as individuals and sometimes friends. On the other hand, it is hard to get on personal terms with something that looks like an artichoke. Add the fact that my study site was cold and inaccessible, and I had a lot of time to myself.

I generally enjoy being alone and have always been able to amuse myself. However, there are limits to this ability. When I began giving plants individual names, I knew I was in trouble. Into this pathetic life came an invasion of four-legged munchkins.

Hyrax are curious animals. One of the few wild animals mentioned in the Bible, hyrax are found in Africa and the Middle East. They are the size and shape of woodchucks and marmots (not quite as big as a bread box), without tails. However, they are hardly related to these rodents. Long suspected of being the closest living relatives of elephants and manatees, they have recently been linked by some authorities to rhinoceroses instead (putting few minds at ease). Having come to appreciate their devious intelligence first hand, I am firmly in the elephant camp.

Rock hyrax live in colonies, and rocky, glacial moraines provide ideal homes and refuges from predators. Both eagles and leopards love hyrax, and the opportunity to duck into the jumble of a boulder field is essential. I chose to erect my camp in the middle of just such a boulder field, which put me in the middle of an active hyrax colony.

Now hyrax can be very active, as well as unusually playful and curious—traits that are more reminiscent of carnivores. Hyrax eat a wide variety of plants. Most herbivores bite off leaves with their front teeth, but as always hyrax are contrary. Instead of normal incisors, hyrax have two bizarre tooth types in the front of their mouths. Below are a set of four teeth, each of which has two narrow slits. Hyrax use these to "comb" through their hair when grooming themselves. Above are two tusk-like teeth up to an inch long. These are weapons that can inflict vicious wounds, as can be attested by visitors who have been fooled by a hyrax's superficially appealing demeanor.

These teeth can be used to pick up objects, but not to bite off leaves. Hyrax must therefore turn their heads on their sides and use their molars, with one eye in the dirt and the other to the heavens, as if on the lookout for a stooping eagle.

Ridiculous as they seem, hyrax are not to be underestimated. Since the arrival of man in their isolated valley, they have become daring and proficient raiders. Using secret entrances to mountain huts, they can clean out a food cache before the mid-morning return of even the fastest climbers.

At my camp there was a colony of about a dozen adult hyrax, with their most recent litters. There was a single male, a battle-scarred old man named Rick. His harem was composed of females of several ages, and my favorite was a young adult that went by the name of Elizabeth. She was always the most curious and adventuresome, and was as cute as a hyrax can be.

I was the first human to live in the territory of this particular colony, and this provided a grace period during which the hyrax were disarming and I was disarmed. It began innocently enough. The hyrax, though wary of me, enjoyed freedom from predation near my tent, making their daily siestas even more relaxed. I thought nothing of leaving my tent open during short absences.

Then food began to disappear. Early on I shamefully suspected both hyrax and the rapacious mountain chats, but never caught anyone in the act. I would return to camp to find all of my neighbors at a respectful distance, but hadn't I had more breakfast leftovers? When confronted, the hyrax always vigorously protested their innocence (or at least vigorously protested). Being a trusting North American in a strange land, I gave them the benefit of the doubt.

Almost every exciting bush story I have heard (or experienced) has been due to the protagonist's stupidity. Smart, careful people needn't be charged by elephants or rhinos; they needn't run out of water, food, or fuel in the middle of nowhere; they needn't shear a Land Cruiser frame (to the amazement of even Safari Rally mechanics); they needn't become the victim of ghastly tropical diseases or unscrupulous locals. The fact that all of these "happened" to me is not because I was exciting, adventurous, or even lucky. They were because I was incompetent (and, of course, a macho kind of guy). My hyrax experiences are no exception.

I was soon overcome by the same urge to befriend wildlife that afflicts every grandmother with a bird feeder. Merely by not having rocks thrown at them, the hyrax became calmer around me than around the park rangers (who knew them better than I did). But this was not good enough. I wanted friends, not just neighbors. So I began giving them food. Leftovers, mainly, and deposited far from the tent. Slowly, I became less careful, and would toss out scraps as I cooked and ate. Sure enough, I made many "friends."

Still, there was the paramount rule: no hyrax in the tent. This was a shame in a way, because it was obvious that they could have made interesting house guests. I didn't need to worry about house-training them; they had their own toilets outside my home and even outside their own. No, the problem was that they were too smart and too insatiable to be trusted around food. They proved capable of getting at all food not under lock and key.

As long as I was within sight of the camp, the hyrax knew that I would come running, delivering a barrage of insults and missiles if they tried anything. However, they soon discovered a singular exception to this rule. I can still remember the first time it happened. I was in the middle of a long visit to my latrine, basking in the glory of the view and the mid-day sun.

Several hyrax were in the area in front of my tent, scavenging the morning's scraps. The one nearest the door looked inside, found me absent, and looked around. Upon discovering my locale, she thought for a moment, and then boldly walked into my tent! She had figured exactly right. I could (and did) scream and shout all I wanted, but was just out of rock-throwing range and in no position to launch a quick counter-attack. By the time I got back to my tent, she was long gone and so were my cookies.

From then on I zipped up my tent whenever I left. Sometimes I would forget or figured I would be back very shortly, but as soon as the hyrax saw my backside hit the seat, the smorgasbord was declared "open." Even when I was sure to zip up the tent, I was always imagining an inexplicable loss of leftovers. These little criminals had me paranoid.

Hyrax are just about the only land mammal I know that do not use their feet as foraging or fighting tools. Primates, bats, rodents, raccoons, and mongooses all have manipulative hands that they use to gather and eat food. Carnivores swat, slash, grab, and dig with their feet. Even ungulates use their hooves as weapons and to dig at food. Elephants, those redoubtable hyrax kin, harvest grass by grabbing a tuft with their trunks and then kicking it free.

Hyrax will have none of that. When encountering an obstacle they face it head on. An overturned cup hiding a piece of potato (during my foolish "feed the cute hyrax" phase) would be inspected from all sides and then knocked over with the hyrax's snout. This snout dexterity had other uses as well. One historic morning, I was reading quietly in my tent when I heard a snuffling at my tent door. Usually I left the doors open while I was home, and hyrax used this as a signal of my presence. On this occasion, the doors were zipped up, and the hyrax were operating on the assumption that I was out counting lobelia leaves.

The snuffling continued and quickly zeroed in on the junction where all the closed zippers met. There was a jiggling of the tent and a little black nose appeared. The nose twitched, smelling the irresistible aroma of stale cookies, and then began to work right and left, up and down. Slowly the zippers began to move. I sat mesmerized. Behind the nose appeared a snout, behind the snout a pair of beady eyes. The whole head was now battering the zippers open. Finally the opening was large enough to admit the hyrax. It was Rick. He stepped quietly into the tent (into my home!) and began walking an obviously well-known path toward my cooking area. But he stopped mid-stride. He looked carefully around and stared right at me. Something was terribly wrong. I was not supposed to be there. But as long as I wasn't moving, he wasn't going to move either. He was betting that either *(a)* I had not seen him, or *(b)* I was not really me, just one of those department store mannequins. There was a long pause. I blinked first. He let out a startled squeak and scurried in panic back through the opening that now seemed barely adequate.

I was outraged. So my leftovers were disappearing after all. I was not (yet) losing my mind. From then on I had to increase my security measures, actually putting a lock on the zippers when I was out. The battle of wits had escalated. I was working on my doctorate, and these guys couldn't even count to ten. It was going to be no contest.

I began keeping a pile of small rocks by my cot. Bopping a hyrax is, as you can imagine, an immensely satisfying achievement, but the hyrax learned so quickly that I rarely got a second shot. So I upped the ante. Instead of merely disallowing hyrax inside the tent, I decided to draw an imaginary line several feet beyond the door, and reclaimed this territory. This led to a renewed but short-lived bout of pebble tossing. These guys learned too quickly for me to alleviate months of pent-up frustration.

Then a friend brought me a long-awaited gift from the States, on order since my early run-ins with my neighbors. It was one of those guns that shoots plastic darts with suction cups on them that are supposed to stick to your little sister's forehead, but never do. I was determined that this ultimate weapon would finally turn the tide.

Unfortunately, by this time all the hyrax had memorized the location of my imaginary line, and would not cross it. I sat with gun in hand for hours, for days, but they never transgressed. The time was not wasted. Hours of practice made me a crack shot. Still they did not put so much as a toe across the line. I took to pretending I was engrossed in a book, hiding the gun between the pages. No go.

Each of us has a breaking point. It may come when your hard disk crashes. It may come after climbing the twelve flights of stairs to the office for research clearance for the twentieth time. It may come when you discover your thousands of individually placed plant identification tags have become local collectors' items. It most often happens in the presence of your mechanic.

For me it came several frustrating weeks after the arrival of the gun, as I watched Elizabeth carefully tip-toe the line between her territory and mine. It became clear to me at that moment that she was taunting me, teasing me with her baleful eyes. There comes a time when the rules of appropriate human conduct do not apply. I slowly lifted the gun, aimed just behind the left shoulder, and fired.

It was a perfect shot. The dart hit true and rebounded a few feet away. Elizabeth let out a little squeak and bounced back herself. She gave me look of outraged disbelief. It was not the new weapon or my accuracy that amazed her. It was the fact that she was hit on her side of the line. I had broken the rules. I managed to suppress a twinge of guilt. I had no conscience. I was in control.

I stared back, smug. She looked from her adversary to the dart and back again. I did the same. She looked long and hard at the dart. So did I. She looked at me again, but this time there was something other than sadness in her eyes. We looked at each other. We looked at the dart. We made our moves simultaneously, but she was quicker and closer. I arrived a half second too late, lunging face-first into the dirt. Elizabeth grabbed the dart in her mouth and took off just ahead of my desperate leap, disappearing into the boulder rubble.

I still had two darts left, but it didn't matter. I was defeated. Morally, intellectually, and strategically, the hyrax had won. Life would never be the same. Luckily, my thesis committee never found out, and a few months later they awarded a doctorate to a guy who couldn't outsmart a pack of five-pound rock-hoppers.

The Hottest Data in Town

by Tim Caro

It was always difficult to buy cooking gas in East Africa in the early 1980s. In Nairobi one had to drive slowly around town examining the Shell Petrol Stations to see whether the long line of empty heavy blue cylinders were still chained to each other outside each cashier's room. If they had disappeared, it meant one of the sporadic shipments had arrived from Mombasa on the coast, cylinders had been filled and retrieved by their owners and, with luck, there might still be some gas available for me to buy. To make matters worse, I worked in a different country, Tanzania, a day's drive and a closed border away. To find gas on one of my three visits per year to Nairobi was luck indeed.

Cooking gas wasn't essential for my work but it made everything a lot easier. First, I could run a fridge off it and this allowed me to keep vegetables for weeks, up to six weeks for tomatoes if you laid them out on the shelves without touching each other, and four weeks for carrots provided they hadn't been washed before they were interred. My battered but reliable four-foot-high fridge could run for two-and-a-half months on a single cylinder if care was taken to turn the flame down as low as possible.

Second, when I had more than one cylinder, I could run a gas stove which meant I could bake bread. I was not an experienced cook, but I had mastered one recipe. When gas arrived, I would take a day off and spend it mixing and kneading dough, interspersing this with oiling and greasing the car, or sorting through data. When everything was ready I would pack the oven with eight loaves in trays borrowed from other scientists. After a single bake and when the loaves had cooled, I would shove as many as possible into the freezer at the top of the fridge, disconnect the stove and store the cylinder for a later date to run the fridge when the current cylinder was finished.

In November 1981, the gas situation was bleak. I hadn't been to Nairobi since July and there were very few scientists working at the Serengeti Wildlife Research Institute at the time, and none of them had crossed the border for supplies in months. Crossing was a very ticklish operation. Tanzania had closed its border with Kenya at the end of the '70s to control tourism. There were ways that residents could cross, but this required a lot of paperwork and friends in the right places. For researchers like me, it was best to get an invitation to a conference in Kenya, even if this only took the form of a good dinner at one of Nairobi's Indian restaurants; but it had to look official. The invitation was then sent with a cover letter from me and the head of the Institute requesting the Principal Secretary in Dar es Salaam to issue a permit to cross the Tanzanian border. Once the permit was in hand, it had to be sent up to Nairobi to the Chief of Police who would issue a permit to cross the Kenyan side of the border as well. Considering one needed to get to Nairobi in the first place for two of these steps, I became adept at thinking very far ahead.

So when my friends, John and Sue, drove up unexpectedly from Dar with bottles of Tanzania gas in green cylinders, not the blue variety from Kenya, and offered one to us it was a welcome bonanza. There was only one thing to do: bake. Actually it came at a good time because my wife, Monique, and I were just about to leave on a long-awaited holiday trip down the west side of the Serengeti Plains where no tourist or scientist visited. Over the last year and a half studying cheetahs in the field, I had often looked to the west at a range of shimmering hills through binoculars and was tempted by its remoteness and beauty. It might seem silly to take time off to explore the Serengeti given that my work following cheetahs entailed long hours of driving, searching, and sitting in the car, but in fact it gave me little time off to visit far-flung areas of the magnificent park. And at this time of year, great herds of wildebeests were amassing in this area. With three loaves of new bread in the back of the Land Rover, and fresh vegetables from Dar, we would have some great campfire meals on vacation. As usual, the baking took longer than expected and I didn't finish until 6:00 P.M. Still keen to leave that evening, we finally locked Monique's car, shut up the house, and headed west to camp at the northern edge of Oldonya Rongai, a range of hills we planned to explore.

Next day, we got up late, but it didn't matter because there was no need to scan for cheetahs at dawn, and Monique, who was monitoring the lion population, was under no obligation in this part of the park to locate and individually recognize lions. Of course we did look out for big cats in a lackadaisical sort of way but only as we meandered south towards our vague goal of the Ndutu Tourist Lodge two days hence, beyond the southern tip of Oldonya Rongai. With tourism in Tanzania at an all-time low, and hardly any biologists in the Park, we had approximately 7,000 square miles to ourselves, as we sat on the roof rack and admired the broad horizon over breakfast. So it was with some surprise when at around 8:00 A.M. we heard the sound of a light plane, clearly getting closer and closer. Soon we could see it, and it swooped down to buzz us about 100 feet off the ground. This was the conventional greeting of airborne to terrestrial scientists, and was always fun so long as you weren't watching nearby cheetahs that hated it. On the second dive, a piece of paper was thrown out of the window; I rushed over and read the message. "Your house has burnt down, return to the Institute immediately; sorry to bring you bad news, Markus."

Now, I had met Markus briefly only a couple of times before, but from talking to him I knew he had a real sense of humor. He lived on Rubondo, an island in Lake Victoria which had recently been made into a national park, and when he visited the Serengeti he would entertain us all with hilarious stories. I didn't know he was passing through the Serengeti that morning. He knew my car, had obviously seen us in the middle of nowhere, and decided to say hello in a manner true to form. It might not have been the way I would greet someone but you could never tell with those Swiss.

During the next two days we had a tremendous break. The cheetahs were too shy to be approached and photographed and it was outside the lion study area, so there was no possibility of taking demographic or behavioral records on either species. Instead we climbed *kopjes,* the outcrops of granite that rise out of the great plains, watched herds of zebras and Thomson's gazelles, and camped out in the most exquisite spots each night sleeping in the bed in the back of the Land Rover. Although this was what I did normally, it was great being a real tourist just for once.

Despite our enthusiasm for camping, on the third morning the thought of a hot shower and slap-up meal at Ndutu Lodge became increasingly appealing. Ndutu, at that time, was a rather rundown tented camp, that was reached by tourists using a road from the east, not via the remote area where we were driving. It had been managed by an ex-hunter for many years but after he had left, his Tanzanian team stayed on and ran the place without supervision. The beer was good and the cooking renowned; the banana crumble was truly outstanding. So cutting the last day short, we headed directly for the Lodge, purposely getting there in time for lunch.

On arriving, we were surprised to find Mike Norton-Griffiths. Mike was a wildlife biologist who had once worked in the Serengeti, but was now based in Nairobi. He had flown across the border in his light aircraft for a day or two to take part in an aerial survey near Lake Eyasi to the south, and had briefly flown into the bush airstrip to take advantage of the Lodge hospitality. We sat down to have a drink with him. Midway through our beers, I looked up to see the Acting Director of the Institute, Mr. Ndugu, his driver, and my Tanzanian research assistant settling down to the bar. This was another surprise since I knew that Mr. Ndugu rarely went outside the confines of the Institute and Park Headquarters because fuel was scarce for Institute personnel. The only time he visited Ndutu was at the end of the month when these three toured the vast Park, reading the rain gauges over a three-day period. My mind raced: This was not the end of the month. Perplexed, I got up and greeted them. The obvious query was "Why are you here?" Silence; no one would answer me. So I asked my assistant directly: "Why are you here?"

"Ask Mr. Ndugu," was the reply.

I did, but it was clear that he was very reluctant to tell me. The reply came slowly, "Did you see Markus?"

"Yes, of sorts, he dipped his wings."

"Did he drop you a note?"

"Yes, he had written some flippant comment."

"Why didn't you come back?"

Silence

"I thought it was a joke."

My answer, when it came, seemed pretty thin, but you had to remember Markus was a joker. His face, all their faces, told me quickly that it was no joke; our house had burnt down. In retrospect, it is amazing we hadn't even considered the possibility that Markus was serious; once on holiday we had become completely carefree and unburdened by the world. But it was his last question that really threw me.

"Why are you running away?"

Mr. Ndugu, it seems, had put two and two together. He had seen Mike, had seen his plane parked outside, seen us drinking with him, and had come to the swift conclusion that we had struck up some deal, heaven knows how, to get Mike to whisk us away to Kenya, since he naturally thought we knew our house was in ruins. It is not uncommon for Tanzanians to disappear and turn up in another part of the country when misfortune strikes. Hence it was not surprising that the notion of us running off would not shake itself from Mr. Ndugu's mind for some time to come.

Like all very bad news, I suppose, it took a little time to really sink in. The details of what happened in the next few hours are somewhat blurred in my memory: anger, sadness, resolve. I can remember driving wearily back along the main road towards the Institute, fifty miles away, trying to think of all the items we had lost. Without question the most important was the data. I had been in the field for twenty-one months, and on each visit to Nairobi, I took the last batch of data, filled-in checksheets, books of cheetah sightings, to be xeroxed. One copy I left in Nairobi, the other I brought back to the Serengeti. But the last visit had been in July, five months ago, and there were no duplicates of anything I had collected since then. Five months' work gone. Incredibly, the photographic card index by which I identified individual cheetahs was in the vehicle; it was the only time I had ever taken it out to the field since arriving in the Serengeti.

Then there were our possessions, everything to make a project run. We had few personal items since we had driven out from London to Nairobi overland, and much of the space in the vehicle had been taken up with Land Rover spare parts. They must have been destroyed too, since the old house that we were assigned on arriving had no outside store in which to put such equipment. The spares alone were worth hundreds of dollars but much worse, were almost completely unavailable in Tanzania. There were also books for the project, the typewriter, tapes, radio, and rugs that we had bought in Nigeria, and all the domestic items for the house: buckets, cooking utensils, and cans of food for the long rainy season just beginning. Because I'm a positive sort of person, we made two lists on that drive home; one for the bare essentials we needed to buy in Nairobi to get the project started again, and the other listing the different groups of cheetahs for which data had been lost, information I would have to collect again.

We returned to a scene of devastation. The old wooden house was razed to the ground and still smoldering four days after the fire. The workers from the Institute were standing around picking through bits of rubble, twisted equipment, molten rubber, and piles of charred notes. We started to hear the stories; everyone had something different to tell. The house had apparently gone up about two hours after we left it and was accompanied by a series of huge explosions. These must have been the petrol cans exploding because without a storage room we had been forced to keep two fifty-gallon drums and a score of four-gallon jerry cans in the sitting room. The explosions were seen from four miles away at the Park Headquarters in Seronera; some apparently thought that the war with Uganda had started up again! Monique's car, on loan from the lion project, would have perished too had it not been for the quick thinking and risk taking by the Institute mechanic, Barnabas. He had broken a window to get in, pulled off the dashboard, touched two of the ignition wires together and backed the car away from the inferno. As it was, the side windows had already cracked from the heat as it had been parked four feet from the house under a tree which was now burnt to the ground. All the lion data and identity cards were still safely inside.

What caused the fire? We have a pretty good idea. From John and Sue, we later learned that many gas cylinders sold in Dar had leaks because the rubber seal at the top partly deteriorated and had not been replaced by the gas company. The most probable scenario was that after I took the green cylinder off the stove at the end of baking, the rubber did not reseal itself and the gas slowly leaked out onto the kitchen floor. Eventually, it would have reached the pilot flame of the fridge ten feet away, with a sufficient concentration to cause an explosion. Once this had happened, the wooden walls and floor, tinder dry after the five-month dry season, must have quickly caught afire. Flames would have raced through the structure rendering it too far gone to salvage, even before the flames reached the petrol store on the other side of the house. A telling gas cylinder with its top neatly blown off corroborated our hypothesis. Incredibly we discovered some hugely bloated petrol jerry cans completely intact but many yards from the house. They must have expanded with the heat, but been hurled away from the fire by the explosions of others, and simply burst open their contents without the fuel igniting.

After surveying the scene and going through the rubble with Barnabas, picking out grilled car spare parts that he thought were still usable, we walked up to the Institute Headquarters to see Mr. Ndugu, who had followed us back from Ndutu. Everyone's nerves were frayed and unfortunately our "discussion" turned into an argument outside the garage. In general, Tanzanians have an eminently sensible custom of keeping the heat turned down in discussions, hating to see arguments develop. Our disagreement hinged on our insistence upon going to Nairobi to buy new provisions and supplies as soon as possible to get the project rolling again. Since such supplies were virtually unobtainable in Tanzania, this seemed like good sense to us but Mr. Ndugu saw it as another attempt to escape an awkward situation, although he didn't say so directly at the time.

That evening we moved our only belongings, sleeping bags, binoculars, camera, and some remaining food from the car into another scientist's house kindly offered to us that night. Just after supper, the lights of a vehicle appeared and drove straight up to the house. Our host went to find out who it was and came back with the chief policeman from Seronera. Mr. Ndugu was with him. Politely, we were informed that our car was going to be taken into police custody for reasons that were not entirely clear. But there was no point in arguing, it simply wasn't worth us being put in jail too, or "lock-up" as it was known locally; all we could do was try to sort it out in the morning. I was instructed to drive the vehicle around to the police station with a car ahead of me and one behind me so that I couldn't make a dash for the bush! I had been thinking fairly quickly since lunchtime which now seemed like a month ago, and it occurred to me that not only had the house gone, but this could be the last I would see of the car, the linchpin of my research in East Africa.

As luck would have it, I discovered we had a flat tire picked up on the way back from Ndutu. I couldn't make the trip to the police station until it was replaced with a spare, so with the aid of a dull torch I eventually managed to change the wheel. Fortunately, the breakdown and the darkness gave me the opportunity to rummage through the well beneath the central passenger seat of the vehicle where all our passports and important documents were always stowed for safety, or so I thought. I stuffed them into a bag, and threw it as far as I could out into the African night. It was the only flat I have ever welcomed in years of fieldwork. Finally, I drove to the station, signed the keys over to John the policeman, and was driven back to the house by our guests.

The next morning in the new light, I retrieved the bag from under a bush, and we wandered up to the Institute on foot, keeping a watchful eye for grazing buffalo that demanded careful respect. There we were greeted by Gaspar, a longstanding research assistant at the Institute, and a national expert on Tanzanian plants. He was so knowledgeable on things botanical that he had been taken to Nyerere's garden to identify all the flowers and shrubs for the President. Gaspar sympathized with our situation and offered us a glimmer of hope: He knew John the policeman well as they both came from the same area on the east shore of Lake Victoria, and would come over with us to the station.

Once we had arrived and sat down, it was clear that John was a much nicer person than I had given him credit for the night before. He too had thought that it was strange that Mr. Ndugu had requested that he confiscate our vehicle, since we had lost so much already, but apparently he was still worried we might run away. Another question was whether the fire had been an accident or arson. It took some time to convince him that nothing but the car was insured, that an enormous amount of work had been lost, and much more. Between Gaspar's Luo, our Swahili and lots of smiles that morning, I think he got the message that we wanted simply the chance to get new supplies and start work all over again. But there was a price for letting us do this: We had to bring John back a generator for grinding maize from Nairobi. Although we knew nothing of this kind of generator, we agreed, for we had no option. John handed the keys back and we promised to report back to him on our return from Kenya. After further discussions and an uneasy truce with Mr. Ndugu, followed up by a formal letter of apology for my arguing with him in public, and after many requests from Tanzanians to bring back baby clothes, sheets, and cloth from Kenya where there was infinite variety compared to Tanzania, we headed north to the frontier a day later.

Two hours into our trip north, we opened up the passport bag in preparation for presenting our documents at the border. Monique's passport was missing. All the car documentation, travelers' checks, research clearance papers, and my passport were there, but not hers. We racked our brains, when had we last taken it out of the car? What were we doing at the time? Working this through we realized that the passport had been left on a shelf in the house a few days before our sojourn to western Serengeti. Now we were trying to cross a border closed to international traffic with no passport. Furthermore, we had none of the necessary permits, not having calculated on this unexpected return to Kenya. But there was no going back after all the sweat we'd been through; either we had to go through the immigration post at the northern tip of the Park and plead mercy, or else veer off into the bush and try to cross Sand River. This was possible, though risky since there were only a few places where the river was low enough to cross. With the river high after the new rains, we were going to have to wing it at the official crossing point.

It wasn't as bad as we thought. The immigration officials had already heard of the house fire on the Park's short-wave radio and were hungry for details. They were also prepared to waive the usual formalities surprisingly quickly considering our circumstances. One never knew how fast things would proceed at the border. If a card game or football match was on, documents were whisked away and stamped in record time. But if it was a slow, uneventful afternoon, a passing resident could become the center of attraction since as few as one vehicle crossed per week, and formalities could bog down to a snail's pace. For doing us the favor of letting us through, customs and immigration put in orders for soap, toothpaste, and cooking oil. We promised to return with these and with a provisional passport from the Dutch embassy for Monique. The Kenyan border ten miles further on took a little more persuasion but because the Tanzanian officials had sent a note with us explaining the situation, they allowed us to pass too.

Six hours later, we arrived in Nairobi, exhausted, bedraggled, and still wearing the same clothes we had on at Ndutu, smeared with charcoal stains from the fire. We walked into our friends' house to an opulent feast and merrymaking, having no clue it was Thanksgiving. Plentiful alcohol and food were good medicine that night, and over the next few days we tried to forget the calamity. Monique helped me buy enough equipment to get started again. We purchased all the items that had been requested from us, except the generator. Maize grinding generators cost thousands of dollars, and were too big even to get in the back of the vehicle. I had no idea whether John knew this or not, but I decided to stall him by telling him

I was still investigating generator prices because the model he wanted was out of stock. While I was there, I fired off letters to my Department and University back in Britain, as well as granting agencies giving me support, asking for emergency money to help me through the disaster. Five days later, while Monique stayed on in Nairobi to start her own project in western Kenya, I started the long journey back to face the music in the Serengeti on my own.

Mr. Ndugu was clearly relieved to see me back, and it convinced him that I was serious about staying on to complete my research in the Serengeti. We got on surprisingly well after that and I was sorry to see him replaced by a new Acting Director a few months later. I was even more sorry when I realized that the new man in charge was going to use my misfortune to benefit his boss.

A couple of months later I was up at the garage fixing the car, when I received a complicated verbal message in Swahili which I could not understand fully. I didn't realize it was a summons from the new Director to discuss a business matter, so instead of following it up, I finished greasing my car, returned home, and then went out to the field again. Within forty-eight hours, a formal letter was delivered which demanded thousands of Tanzanian shillings to recompense the Institute for my negligence in allowing the house to catch fire. The amount was simply unpayable. It was more, far more, than I had in my research or personal accounts. I was furious and worried; the fire had been terrible, but this doubled the insult. What should I do? I wanted to go on with my research and stay in Tanzania but I couldn't pay this bill. I decided to seek the advice of other Tanzanians at the Institute. Some said I should get a letter from John the Policeman saying that the fire was an "Act of God," another said pay in black market shillings and it won't be so bad, and a third said "Go and bargain with the director, this is only a starting price." I did all three. With the help of John's letter, I argued that the demand was quite unreasonable, but that I sympathized with the Institute losing property. In the end, I pulled the figure down to half the asking price, and at black market rates the amount came exactly to the sum total of what I had been sent by the University and granting body for emergency support. It was rough justice but had a certain balance to it.

Of course, I eventually made up the months of lost data by observing different cheetahs matched for the same age and sex, and I am still working in Tanzania. Yet lost data have a ghostly quality. One can partially remember important incidents and even the act of writing them down, but the records can never be scrutinized and used in statistical analyses that are a critical component to my work. With only memory to rely on and no way of checking the facts, it is all too easy to inflate their importance and imagine that they were the hottest data in town.

Section IV: Clash of Cultures

A Trip to the Che-Wong

by Elizabeth L. Bennett

WE HAD BEEN walking through the hot, steamy rain forest for about three hours when we eventually came to a large clearing. After unloading the heavy backpacks I pottered down to the riverbank with the others to splash my face and arms with some refreshing cold water. One by one, we then climbed up into the wooden hide, made by the Wildlife Department for spotting gaur and other large mammals. At this hour we didn't expect to see anything, so we slugged back some water from the drinking bottles, slowly redistributed the backpacks between us, and moved on.

This was 1981, and I was with three other English wildlife biologists and three Orang Asli guides in the Krau Game Reserve in Peninsular Malaysia. I had already been working in the Reserve for two years, studying the ecology of the banded langurs, a type of forest-dwelling monkey, as part of a larger program between the University of Cambridge and three of Malaysia's universities. On this particular trip, however, monkeys were not foremost in our minds. We were on our way, via a ten-hour trek, to visit a semi-nomadic group of forest dwellers.

These people were Orang Asli, the aborigines of Peninsular Malaysia. Although originally forest dwellers, many Orang Asli had become settled farmers. There are many different tribes of

Orang Asli in Peninsular Malaysia, of which the two in the vicinity of the Krau Reserve were the Jah-Hut and the Che-Wong. We were heading toward the last remaining village of the forest dwelling Che-Wong. Because I had been quite busy before the trip, I hadn't found out a great deal about them, except that there were only about twenty adults in the village. They made small clearings in the forest where they lived, grew a few vegetables, and hunted for meat. After several years, they would move on to clear another small area and continue living in the same way. Our guides belonged to the Jah-Hut who were living as settled farmers but were still close to the forest and depended on it for wood, rattan, fruits and, occasionally, wild meat. The Jah-Hut are also famous for their splendid carvings of the many spirits which they believe inhabit the forest, rivers, and farmlands.

It was this fascination for the Jah-Hut which had prompted our trip. During the previous two years, all of us had come in contact with the Jah-Hut; they had been our guides and mentors in the forest, and we had developed a great respect for their gentle nature, culture, humor, and knowledge of the forest. We were keen to meet and learn more about their nomadic, forest-dwelling neighbors. So although this trip involved trekking through the forest, we all regarded it as a holiday and viewed it with excited anticipation as a break from the usual rigors and stresses of fieldwork.

About an hour after that initial water stop, Clive suddenly swore and announced that he had left his binoculars back at the hide. The Orang Asli guides understood the problem and presumably realized that if they returned for the binoculars it would be far quicker than if we rather ungainly Europeans did. Being extremely polite, they didn't phrase it that way; they merely offered to go, which seemed the best solution all around. While they headed back, the rest of us searched for a suitable spot to stop for lunch and wait for them. Included in our party was Kalang anak Tot, a Jah-Hut renowned amongst fieldworkers in that part of the world for his exceptional knowledge of the forest and extreme good humor. He had been working with me for the past two years and had not only taught me much about the rain forest, but was the first person to show me how to enjoy its intricacies and beauty. He was obviously fascinated by the forest and so completely at home in it that when I first arrived and we went into the forest together to look for monkeys and identify trees, he was constantly pointing out other little things that interested or amused him.

"Look at the trail of termites," he would say, putting a twig across the trail to cause a great traffic jam as the termites suddenly could no longer "smell" their nest mates' trail in front, so went round in all directions as the great backlog of others all built up, somewhat reminiscent of New York during rush hour.

Kalang was in the lead and had just crossed over a small stream when he stopped to indicate that there was something ahead. Clive, right behind him, was obviously excited about what was there. Unfortunately, I was near the back, and because the bridge comprised a single log balanced above the stream, it took me some time to cross with the heavy pack on my back. I heard some dogs barking, but by the time I reached the far bank, I missed viewing some Asian wild dogs up ahead—very exciting because they were shy and rare in this part of the world.

About half an hour later, we came to a lovely little beach by the river which was an ideal place for lunch. Everybody splashed cold water on themselves or plunged completely in the river before settling down.

"Okay, who's got the pack with all the food in it?" someone asked.

We had divided the various goods between us before starting so that nobody had to carry a pack continuously and we could swap them to allow time for cooling off; it's jolly hot walking in the heat and humidity with a heavy pack on. We all started rummaging through the packs, initially just taking a few clothes off the top, then pulling things out with more speed as we delved for the bread and cans of curry. Gradually, it became apparent that the one backpack with all the canned food, bread, and rice in it must be sitting on the floor back at our starting point. By then, we had traveled too far to make it back before dark. At this realization that this normally highly organized team from Cambridge had entered the forest for three days but left all the food behind, we started giggling, then collapsed on the beach with contagious laughter. Kalang, who didn't understand English, looked bemused and suspiciously asked what we were laughing at. When we told him, he obviously couldn't believe that these English lunatics *(a)* had forgotten everybody's food (including his), and *(b)* were actually laughing at it! He gave a subdued and hesitant giggle since that seemed to be the thing to do, but continued to puzzle over our uncontrolled mirth. Luckily we did have with us a bag of oranges which we proceeded to feast upon with our supply of drinking water. When the other two Orang Asli guides joined us, clutching the lost binoculars, we knew that our only option was to carry on toward the Che-Wong village. We checked with Kalang to make sure the Che-Wong would have enough spare food for us. When he assured us that they would, we continued on our increasingly weary way.

Suddenly, Kalang looked wary again, taking off his conspicuous white hat and tucking it into his pack.

"Elephants," he murmured. These were the only animals in the forest of which the Orang Asli are frightened. Kalang had spotted the tracks of a female cow and calf, so we walked quietly for a while in case they were nearby. When crossing a small streambed, we saw the enormous tracks of a tiger, presumably following the elephants in search of a meal.

Mike's watch bleeped the half hour, so we all exchanged backpacks. By this time it was late afternoon and I was starting to feel that if we had to go much further, my legs wouldn't make it. Luckily, there soon appeared a clearing. I wouldn't have to lose face by asking for another rest after all. The clearing comprised a mass of felled tree trunks with tapioca and other fruits and vegetables planted untidily amongst them. Over the far side was one large wood and bark house on stilts, with a palm-thatched roof and smoke filtering through it into the air above. The smell of wood smoke penetrated the clearing. We tottered across the fallen tree trunks towards the house, then stood around while Kalang went in to explain who we were and about our unfortunate food predicament. In the meantime, we watched the small children playing, including a kid who couldn't have been more than four years old smoking the most enormous cheroot as he peered at us from behind the house steps.

Kalang came out smiling and said that we were going to sleep in the grain store, a small hut on low stilts across the clearing which was dry and just big enough for all of us. We would go into the communal main house for our meals. The Che-Wong had plenty of rice which they were happy to share with us in exchange for money and other goods such as T-shirts.

After a bath in the river, with the pleasant sound of hill mynah birds whistling above, several of the Che-Wong visited our hut for a chat before dinner. Fortunately, they all spoke Malay which gave us a common language; some of the older women speak little apart from Che-Wong. We spent a pleasant half-hour as it was getting dark going through a word list; a friend at the University of Malaysia had explained that a list of about 100 key words in a language can give a good idea about its derivations and relationships to other languages. So I ran through as many of these as I could, asking for the Che-Wong word, with numerous good-humored breaks for explanations in a combination of Malay, Jah-Hut and Che-Wong between us to get across particularly evasive meanings.

When darkness settled, the eldest of the Che-Wong, Beng, invited us into the communal village house for dinner. Beng had short, graying hair and an aged, wrinkled face. In fact, he was probably much younger than he looked; life expectancy amongst the Orang Asli is generally much lower than amongst most Malaysians. Even in Kalang's more settled village which was close to a clinic, the oldest inhabitant by far was only about sixty. On that basis, we surmised that Beng might only be in his mid-forties. He led us, clutching torches, back across the felled logs to the main house. We left our shoes on the ground before climbing up the wooden steps into the house where on one side of the room was a large stone grate with a fir and Beng's wife was cooking. Like the other women, Beng's wife was dressed with only a small sarong round her waist forming a knee-length skirt. The men were more modern looking in their soccer shorts and T-shirts. We sat on the floor in the smoky room, lit by small kerosene wick lamps. In the flickering dim light, we caught glimpses of fish nets, blowpipes, and trophies from hunted rhinoceros hornbills hanging in the corners of the room. After a short period of conversation, fish soup plus large piles of hill rice and baked tapioca roots, ideal after the long walk, were placed in front of us. The Che-Wong hadn't been hunting for a while so the food was flavored with burning hot "padi chilies."

After we had eaten our fill, we settled down against the walls with our mugs of boiled water for the evening's entertainment. This was our first night deep in the forest. What would the evening have in store? There was obviously some excitement afoot. Were they going to give us a display of dancing, gongs, and drums? Or indulge in some ritual to greet the arrival of strangers? I knew that the forest people of Sarawak in East Malaysia had many elaborate ceremonies when visitors arrived, so maybe the Che-Wong did something similar. Or were they going to recount tales of the forest, its animals, and spirits? The history of the village and its people? Ah, they were obviously all settled and ready. One of the younger men, Langit, moved across the room, leaning low in courtesy when he walked directly in front of anybody, and ceremoniously removed the cloth cover over . . . a television! Apparently, they had just recently acquired it. It ran off a car battery, and somebody had to take a long walk once a week to get it recharged. We spent our first evening with these true forest dwellers, in the middle of the jungle, watching "The Waltons"! In spite of the fact that they couldn't understand the language, or that the cultural context must have been incomprehensible, everybody in the room was riveted to the screen, completely absorbed in the program. Even the commercials about road safety and the use of pedestrian crossings held their attention.

Eventually, we excused ourselves and returned to the hut. Crossing the fallen tree trunks to get there in the dark wasn't easy, and Mike slipped off and twisted his ankle, provoking the most appalling outburst of foul language. Afterwards, he explained it was because he was worried he wouldn't be able to walk back out of the forest again. (Images of long periods of a protein-deficient diet and "The Waltons" looming before him?) We staggered into the hut where we got out our sleeping bags and laid down, squeezed in a little like sardines, on the bark floor.

The next morning, after a breakfast of rice and padi chilies, we set off with Langit, several other Che-Wong, plus our own three Jah-Hut to *Lata Tujoh,* meaning seven waterfalls. It was about an hour's walk from the village, and we were in a much better state than the previous day to appreciate the forest. I have clear memories of hearing a pair of rhinoceros hornbills honking away in the distance, and of coming across a tree which the Che-Wong said they had felled for its bark which was particularly good for house walls. We also walked right past a really beautiful red-tailed racer, a bright green snake about six feet long with a red tail and clear glassy eyes. These snakes are probably quite common but so enigmatic that they are rarely seen. I wouldn't have spotted this one, even though it was only two feet from the path, until Kalang with his sharp eyes pointed it out.

We finally arrived at Lata Tujoh, one of the most unspoiled and beautiful places I've ever visited. It seemed that no humans had ever been here before. We knew that they had, but there was no rubbish or other indications of human presence. These were the headwaters of the Lompat River, which comprised a series of waterfalls with large, deep pools between, surrounded by the tall primary rain forest. Clive immediately brought out his fishing rod, carried painstakingly all the way up here, and perched himself at the bottom of one of the waterfalls. The rest of us pottered around the edge of the forest and swam in the deep pools. After about an hour, Clive still hadn't caught anything. Langit casually produced a sort of harpoon, made from a piece of straightened coat-hanger wire with a length of an old rubber inner tube cut into a strip and attached to one end. He put on a diving face mask and dove into the deepest pool. Two minutes later he resurfaced with a broad grin, holding up a large fish. The triumph of simple over sophisticated technology! After a while he had enough fish to merit a barbecue. He and Kalang started a small fire on the rock next to the river and barbecued the large, delicious fish—a welcome addition to the rice and chilies. After lunch, the Orang Asli showed us that one of the waterfalls made a great slide, with a piece of bark used as a toboggan to zoom down through the water between pools. A short spell of this, then it was time to dry off and wend our way back to the village.

Another evening of stories ensued, during which the Che-Wong started chatting about how they were unhappy with the Government's plans to resettle them out of the forest where they would become farmers. They felt that this would remove them from their culture and background which lay in the forest. In fact, we got the impression that they had somewhat given up; they had not been hunting recently, some of the children looked distinctly malnourished with potbellies, and we felt that the old way of life was soon to be lost. This is a real dilemma facing governments and forest people throughout the region: How to give such people the benefits of medicine and education, without forcing it on them, destroying their culture, and removing their choice of how they would like to live? There are no easy answers, but it was certainly clear that the Che-Wong were unsettled about the changes they foresaw. A little more all-entrancing TV, then back to our bark floor for an early night to build up strength before the long walk out of the forest.

The next morning, we woke and piled out of our hut in an extreme hurry when Andy noticed that we were sharing our tiny room with the most enormous spider. As the cool dawn mist still hung through the trees, we bolted a quick plateful of rice. Beng's wife kindly handed each of us another large bundle of rice wrapped up in a banana leaf for the journey home. In exchange, we paid for their generous hospitality, and both parties wished each other well before we waved good-bye and set off behind Kalang's brisk pace into the forest. Even though our packs were just as heavy, the walk back seemed somewhat easier than on the way in. The only person for whom this cannot have been the case was Kalang himself. Soon after we had started, he stopped to examine a large, flat stone which must have weighed a good fifteen pounds. He picked it up, felt it carefully, and said that it would be an ideal sharpening stone. I was just assuming that he was regretting he had found it so far from home so he couldn't use it, when he opened up his own backpack, emptied out his clothes and put the stone at the bottom before piling everything back in, slinging the pack on his back, and walking as easily as before! It was certainly a lesson in different people's priorities; we had managed okay without our cans of food, and the fishing rod was useless. On the other hand, I wouldn't walk for a full day every week to charge a battery merely to watch television, and certainly wouldn't lug a heavy stone for ten miles, even if I was in good shape!

We recognized the trail about a mile from camp because it was on the edge of my very familiar study area. In spite of our exhaustion, Andy and I decided to race the last stretch to the rest house. We belted out of the forest, panting and staggering and laughing hysterically, to arrive filthy and hot, looking as though we must have had a tiger on our tails. As it turned out, we ran straight into some smart and unfamiliar Europeans with cameras—a team from the BBC who had come to film the gibbons of the area! Sigh—so much for our image as serious field biologists! In the meantime, it was into the river for a bath, then we rummaged through our abandoned food backpack for a can of mutton curry and some slightly limp vegetables—bliss!

Social Anthropology at the Emali Hotel

by Dorothy L. Cheney

The Amboseli Ranger Post in the village of Ol Tukai was a small wooden shack situated opposite a long line of parked lorries and four-wheel drive vehicles. Each vehicle had been painstakingly painted with an elaborate seal which, together with a brief message, announced an uneasy alliance between Kenya's Ministry of Tourism and Wildlife and some international aid organization. And, in keeping with the courteous suspicion common to arranged marriages, the seals left little doubt about the form or direction of the dowry: "A gift to the people of Kenya from. . . ." Then would follow an acronym representing some international organization typically based in Switzerland, whose generosity was tempered only by the high salaries of its officials and the fabulous interiors of their offices.

The *largesse* was literally on display, for none of the vehicles ever worked. Each had lost some crucial part at the inventive hands of local drivers and the Ol Tukai workshop crew, whose proud motto, "If we can't weld it, it can't be welded," resounded with disturbing assurance through the workshop. Here rusted a road grader, whose axle had been sacrificed in a valiant effort to break the land speed record between Ol Tukai and Kilunyet, the Amboseli Park Headquarters seven miles away. Next to it languished a tipper, victim of the Amboseli football team's attempt at a triumphal midnight return from the nearby town of Oloitokitok without benefit of battery or headlights. Several Land Rovers and lorries had succumbed, following weeks of torture and the random removal of their intestines. One of the Land Rovers had been spirited enough to attempt a reprisal against Lodo, the crazed Maasai mechanic who had briefly held sway in the repair shop; it had rolled over Lodo while he gouged at its undercarriage with a pointed stick. It too had eventually conceded defeat, however, and now stood, all movable parts securely welded, at one end of the line.

In fact, only one vehicle in the park's fleet could remotely be described as being in working order, and it had been requisitioned by the warden for what were universally referred to as "lodge patrols." There had been conscientious wardens in Amboseli's past, who had been visible and accessible at all hours of the day. The current warden, however, clearly felt that an aloof demeanor enhanced the dignity of his position, and he typically remained quiescent until around five o'clock in the evening, when he would drive his Land Rover ("A gift from the Conservation Club of Sweden. . . .") among the park's lodges and personally verify that there was an ample supply of beer and spirits at the bars for the park's tourists.

This left the rangers with no "transport" and therefore no apparent means to patrol the park. They spoke vaguely of foot patrols, which had been carried out at some point in the remote past and would be conducted again, if all went well, at an unspecified time in the future. But foot patrols required some planning and had not yet been mandated by those authorized to do so. In the meantime the rangers waited.

But there was much to discuss, and when we researchers had nothing better to do than analyze what too often seemed like arcane data, we would sometimes stroll over to the ranger hut to join in whatever conversation held the floor at the time.

On one languid afternoon, for example, Kasaini, a ranger who had recently returned from a visit to his wife in Oloitokitok, reported that travelers were again being waylaid by the *watu wa damu* (literally, "people of blood"), medical technicians from Tanzania who held up *matatus* (bus taxis) along the road, killed all the passengers, and then inscrutably drained their blood for later sale to Europe. They left the dried corpses, like prunes, scattered along the road. After pondering this news, the rangers turned the discussion to the relative singing merits of Madonna and Yvonne Chaka Chaka, the renowned South African disco queen, whose beer-drinking song, "Uqomboti," was sweeping the tour bus drivers' camp bar.

From there, the conversation flowed effortlessly to the size of European women's breasts, the circumference of the Earth, glasnost, and the violent, though in retrospect highly predictable, breakup of the longstanding affair between Patrick, a receptionist at Amboseli Lodge, and Nellie, the County Council radio operator. Nellie, it seemed, had finally become suspicious of Patrick, whose clandestine tryst with Esther, another receptionist, was common knowledge to most of Ol Tukai. Nellie had confronted Esther, whose response had been to "dent" Nellie's forehead with a *panga* (machete). Now there was some concern about whether or not Nellie, whose tribe put some reliance on witchcraft, would try to put a spell on Esther, whose tribe did not. The metaphysical ramifications of this problem produced the only lull in conversation that afternoon.

We researchers were not easily accommodated into the Ol Tukai chain of being. Not only were our vehicles new and in infuriatingly good working order, but we drove around the park as if it were our private domain, talking ingenuously about "our" animals and "our" study areas. We were also ridiculously eager to offer lifts to the rangers and their families, and although we scrupulously avoided any hint of condescension, our relative wealth inevitably imbued our offers with a suggestion of *noblesse oblige.* Nevertheless, our offers were always accepted, if only because, unlike the tour drivers, we never dared to ask for payment in return. We were more than happy to take passengers with us when we drove around the park engaged in our esoteric projects, and we even made unscheduled stops along the way. And, to the mystification of the residents of Ol Tukai, who strove to adopt western customs and values, we seemed to derive particular satisfaction from helping the traditional, even "primitive," Maasai tribesmen.

The Maasai who grazed their cattle in the dry scrubland surrounding Amboseli often walked into Ol Tukai to buy cooking oil, tea, and sugar in the dry-goods store. Usually it was the women who walked in, carrying calabashes of milk to sell to the staff at the tourist lodges in return for cash to spend at the store. Even though the prices at the store were fixed according to government standards, the negotiations for their purchases often took up the better part of the morning. This was because many of the Maasai still adhered to the colonial monetary standard of twenty shillings to the pound, fully two decades after Kenya had officially converted to the more comprehensible decimal system in which a shilling was simply a shilling and not subject to any other hierarchical transformation. Given the fact that the Maasai had adopted few, if any other, British customs, it seemed an arbitrary and awkward choice. To make matters worse, the Maasai referred to pounds as rupees, and had added an enigmatic decimal to their calculations. As a result, a can of cooking oil, whose price could ordinarily be stated with simple elegance as "twenty shillings" would be translated as "ten rupees," while a one hundred shilling note, with the digits "100" emblazoned unambiguously on both of its sides, would be tendered as fifty rupees. Njeri, the young Kikuyu woman who ran the dry-goods store, could adjust to most such conversions with some aplomb, but when business was brisk she occasionally forgot to make the necessary calculation. In the heat of a sale, she might mistakenly give the price of a forty shilling can of cooking oil as forty rupees, a price that would be received with skepticism by her clients, who would recognize immediately that the price was double the government standard. Njeri's knowledge of the Maasai language was limited to a few trite greetings, and since her attempts to mollify her clients' suspicions relied upon abstract algebraic conversions, even the simplest transactions often took up the better part of the morning.

Their tortuous purchases completed, the women would gossip at the store and perhaps try on the reflector sunglasses owned by some of the ostentatiously blasé lodge staff. Then, in the late afternoon, they would pick up their goods and begin the dusty walk back to their *manyattas*, or settlements. There was an air of excitement to the walk, not because of the possibility of encountering elephants or other animals, since it was impossible to walk anywhere in Amboseli and *not* see wild animals, at least at a distance. Rather, it was the generally unrequited potential of a lift. When the cloud of dust heralding the approach of a vehicle first appeared on the horizon, the Maasai would begin a languid limp-wristed wave which, though lacking a handkerchief, was faintly reminiscent of a departure from an Edwardian train station. The women were more or less reconciled to being left in the wake of the receding bus, swallowing mouthfuls of dust. The rare man who had made the trek into Ol Tukai, however, generally received this rebuke with some outrage since, unless he was accompanied by a goat, there could be no reason for failing to give him a lift.

Maasai men, particularly the warriors, did not presume to be snubbed. They were God's people, and they expected to be treated with respect.

When Rick first arrived in Amboseli to begin his study of elephants, he made it a point to give lifts to the Maasai he met along the road, and since he covered a large distance every day on his surveys he was often willing to take people not just along the road for a few miles, but across the bush and directly to their manyattas. One of the first Maasai who arranged for Rick to befriend him was Kerasoi, a less than prepossessing warrior with one eye and huge protruding teeth. Kerasoi came from a very traditional family, and he was the only member of his family to speak Swahili or to travel regularly to Ol Tukai.

The role of the warrior in today's Kenya is difficult to define. In the past, it had been the warrior's purview to raid cattle and avenge the inevitable retaliatory raids. As a result, warriors neither tended cattle, a chore left to their younger siblings, nor owned cattle, a privilege accorded their fathers, but instead spent their time in a constant anticipatory state of war. Now, however, the government frowned upon these activities, and the warriors were forced into an even more detached state, preparing for the anticipation of possible raids. To alleviate the stresses of this purgatory, the warriors frequently walked into Ol Tukai, to drink sodas and observe the bizarre dress of the tourists. It was on a contemplative stroll back from the Amboseli Lodge Bar that Rick picked up Kerasoi.

Rick agreed to drive Kerasoi directly to his manyatta and asked him for directions. Kerasoi tossed his spear into the back of the car, climbed in next to Rick, and pointed out a distant tree on the horizon. Rick drove his Suzuki off the road and into the rough, thorn bush country that rose from a dry lake bed and headed in the direction of the tree. They drove silently for fifteen minutes as the tree receded into the distance and disappeared, only to be replaced at the next rise by three completely new trees, teasingly silhouetted against the horizon.

"My home is just there, next to that middle tree. You just drive there and you will find it," said Kerasoi in reassuring tones. Rick drove on for another twenty minutes, and the trees remained tantalizingly small. Rick was just beginning to wonder if the Maasai had taken up the art of Bonsai when Kerasoi abruptly announced that they had reached his manyatta, a circle of brown, low-lying huts that blended in well with the barren landscape. The Suzuki was suddenly enveloped in a cloud of dust from a passing herd of goats and Kerasoi leapt from the car and strode away. Rick turned his car around and bounced back for forty-five minutes through the deepening twilight until he found the dirt road.

And so it went for the next two years. Kerasoi had an uncanny ability to appear, like a genie, in even the remotest areas of the park; and he seemed to time his appearances to coincide with those moments when Rick, in misplaced maudlin fervor, was reclining in his car gazing at a sunset or at elephants backlit against a vermilion Kilimanjaro. Before his mood even had a chance to break, he would be bouncing through the bush to a tree that lay "just there," though never backlit, against the horizon. Rick consoled himself with the thought that he was experiencing a side of the untamed African savannah that few tourists ever encountered, although it also had to be admitted that few would have sought to do so. For his part, Kerasoi never thanked Rick for his livery services, something which galled Rick even though his more detached anthropological muse told him that, in the untamed African savannah, no Maasai had ever thanked a non-Maasai for anything.

Finally, Rick cracked. One twilight evening, as he and Kerasoi bounced over the rough terrain, away from the sunset and Kilimanjaro toward the obligatory tree dancing like a hypnotist's pendulum on the horizon, Rick began the speech that he had composed during several previous rides.

"Kerasoi. It has been many months now since you and I became friends, and since then I have given you many lifts to your manyatta." Rick paused. His Swahili was fairly rudimentary and it had evolved a flowery, formal tone whose origins Rick was certain had nothing to do with him. What he had wanted to say was "There's no such thing as a free lunch," but the only way that he could think of saying this translated roughly into a rambling treatise on supply side economics. Instead, Rick blurted out a sentence that he was reasonably certain approximated, "What have you done for me lately?"

Kerasoi looked shocked.

"Rick," he finally exclaimed; there was a baroque elegance to his prose that put Rick's crude subject-verb-noun phrases to shame. "For these many months I have known you as my friend. How could you conceive that I would accept these small favors from you without some gift from me? In actual fact, I have given you a cow." The words hung for several seconds in the dusty cab of the Suzuki. A cow? Rick knew very well that in the Maasai firmament cattle occupied a position far loftier than money, and though this position might be less celestial than that of children, the cow's place relative to wives was a matter that would best not be put to test. On the other hand, the cow, rather like the tree on the horizon, had never actually materialized.

Rick thanked Kerasoi.

"I would like to see this cow that you have given me." Kerasoi looked crestfallen and explained that, much as he would like to show the cow to Rick, she was not presently at the manyatta toward which Rick was inexorably heading. Instead, she was grazing at another of his father's manyattas, under the care of one of his father's other wives.

"However," Kerasoi reassured Rick, "She is young and very beautiful and very black."

Although the Maasai love their cattle, and believe with supreme confidence that all of the cattle in the world belong to them, they also acknowledge the necessity of ready cash, and occasionally sell some of their cattle at markets. One day, Kerasoi mentioned to Rick that he and his father had decided to sell two cows at the market in Emali, a town some thirty miles from Amboseli, and that it just so happened that one of the cows they were going to sell was Rick's. Rick had long since accustomed himself to owning an intangible asset that made no demands on him whatsoever, and he was eager to accompany them. After all, he reasoned, it's not every day that you can experience a way of life that is fast disappearing—not every day that you can trek over endless plains with people who really feel at home in the bush, and who think nothing of sleeping outside in the midst of elephants and lions.

It was Rick's idea to walk with Kerasoi to the Emali market, a trip that would take one-and-a-half days, since the cows could not be walked fast if they were to be in good condition for the sale. They would sleep outside along the way, and carry only spears and whatever else the Maasai normally take with them. Kerasoi was dubious. He agreed, reluctantly, that since one of the cows was Rick's he had every right, and even perhaps a duty, to attend the sale, but he was puzzled by Rick's desire to walk to Emali. Generally speaking, the cows were walked to market by young herdsboys, while Kerasoi took a matatu. The romantic allure of a walk over dusty bush country escaped Kerasoi entirely. Why walk when you could ride and arrive there in two hours? Nevertheless, Kerasoi was by now inured to some of the romantic delusions of white people, and he reluctantly agreed to walk with Rick. The herdsboys would be mystified but delighted.

Rick too had some questions. It was one thing to trek through Maasailand like the nineteenth-century explorer Joseph Thompson, at one with the untamed African savannah and its people, but why go to Emali? Why not walk to Namanga, a much more impressive town which was half a day's journey closer to Amboseli and had a cattle market identical to Emali's? On this point, however, Kerasoi proved adamant. The cows simply had to be sold in Emali.

As the day of Rick's Rousseauian adventure approached, a heated debate ensued with his wife, Phyllis. In order for his trip to have any authenticity, Rick argued, it was essential that he travel as a Maasai would, without benefit of western accoutrements or creature comforts. This meant that Rick would undertake his trek with no change of clothing, no soap, no deodorant, and no toilet paper. Phyllis, who had arranged to pick Rick up in Emali with her car following the cattle sale, balked. It was all very well for him to be one with nature, but it was unfair to subject her car's interior to this experience. She insisted that Rick take a small backpack with deodorant and toilet paper. Rick imagined himself tiptoeing surreptitiously into the bushes, pink toilet roll in hand (at that time Kenya's Rose toilet paper company had a surfeit of pink dye). He would be a laughingstock. Nevertheless, Rick was certain that after two days of tacking to market in the wake of an ambling, grazing cow, a walk back to Amboseli from Emali would hold limited appeal. In the end, Phyllis ceded the deodorant, and Rick capitulated to the toilet paper.

On the appointed morning, Kerasoi appeared in Ol Tukai, ready to direct Rick to his father's manyatta. It had been agreed that Phyllis would drive Rick and Kerasoi to their departure point, for, as Kerasoi cogently argued, it would be absurd to start walking before it was absolutely necessary to do so. As they drove off the road toward the inevitable tree on the horizon, Kerasoi explained to Rick that although his father spoke no Swahili he would be glad to translate any messages of greeting that Rick might have for him.

Kerasoi's father emerged from the manyatta in the obligatory British army greatcoat that many Maasai elders wore, apparently with little regard to climatic conditions. Unlike most other old men, however, Kerasoi's father wore no woolly hat. Instead, the top of his head was daubed with a large "X" sculpted from what appeared to be mud, which graced his skull like a pimento atop an avocado. Solemnly, the old man greeted Rick and welcomed him to his manyatta. An awkward conversation ensued. It rapidly became clear, to the father as much as to Rick, that, beyond the most preliminary of conversational gambits, there really was very little to say. Rick complemented Kerasoi's father on the fatness of his goats. The father groped for a suitable riposte, since there was little in the way of earthly possessions, beyond his very unimpressive backpack, to recommend Rick. Kerasoi's translations were painstakingly accurate and lengthy and contributed little to the conversational flow. At last, the old man attempted an almost arbitrary stab: "No doubt you have noticed the goat shit on my head?" Rick was nonplused. Would it be a social gaffe to admit that the "X" had not escaped his attention? Fortunately, the old man seemed to recognize the awkwardness of the moment. Before Rick had the opportunity to reply, he declared with some finality, "It is there because I have a headache." Rick could add little to this incontestable conversation stopper, and it was decided that he and Kerasoi would set off immediately for Emali.

Rick's cow was indeed black, though rather small and scrawny. She stood like a shadow at noontime next to Kerasoi's rather grander heifer. No matter; she could walk, though slowly.

It quickly became apparent that there was limited appeal to following a creature that literally ruminated as she strolled. The sun was hot, Rick was thirsty, and the nuances of his secret supply of toilet paper weighed heavily on his mind. Matatus flew by in clouds of dust on the horizon, only to disappear as mirages at the next rise. Social anthropology was a ridiculous profession, Rick decided. What he was doing amounted simply to demeaning voyeurism, particularly since it was unclear who was observing whom. Kerasoi hummed and strolled easily along. What was Kerasoi so happy about? Hadn't he been reluctant to walk in the first place?

That night, as Rick lay shivering on a pointed rock that his clearly inadequate supply of toilet paper did little to pillow, Kerasoi sighed happily and said, "Tomorrow by this time we will have sold our cows and we will be sleeping at the Emali Hotel." This statement failed to have a reassuring effect. Granted, Rick had never entered the Emali Hotel, but its exterior of rickety tin walls and sunbleached roof promised little to the weary traveler other than a too literal sultriness.

"What," Rick asked again, "is so special about Emali? By now, we could have been in Namanga, which has a very nice hotel and a disco. From what I have seen, the Emali Hotel has only three small, hot rooms made of tin."

"What you say about Namanga is true," replied Kerasoi. "However, Emali has one thing that Namanga does not have."

Here, Kerasoi paused, as if to emphasize the magnitude of the revelation to come. "Emali has a prostitute that does it on her back. Good night, Rick."

Rick was hit by a sudden frisson. Had the missionaries had as little effect on the Maasai as the Maasai liked to claim? To what extent should an anthropologist take cultural relativism? And what—dared he ask?—was the preferred method? He gazed up at what now seemed to be a brazen display of stars and fell asleep.

Emali proved to be, figuratively at least, an anticlimax. Rick's cow was met with an apathy bordering on ridicule. Kerasoi did not have to reject the low price that he had claimed would be an insult, because no offer for her was made. She was sent back to Kerasoi's father's manyatta with a small herdsboy, both Kerasoi and Rick now of the opinion that they would benefit from a matatu ride. Emali boasted few attractions other than the post office, a petrol station, and a few dry-goods stores, and after a lackluster meal of goat stew and a few warm Tusker beers in the Emali Hotel saloon, Kerasoi set off to seek his destiny. Since there seemed to be an element of déjà vu to Kerasoi's choice of evening activity, there was little left for Rick to do but retire to his single cot in Room 2. In spite of the considerable activity taking place in Rooms 1 and 3, Rick fell instantly asleep, his diminished roll of toilet paper still pillowing his head.

Rick was awakened suddenly a few minutes after midnight, as Kerasoi stumbled into the room followed by three other Maasai, all of them slightly the worse for wear.

"The prostitute has moved to Namanga, and so it is time to sleep. These are my best friends," he said, introducing Rick to three glazed, impassive expressions. "They are overly tired, and since there is no room for them in the other rooms, I have told them that they can stay with us." Rick must have revealed some consternation, for Kerasoi added hastily. "Don't worry; they will sleep on the floor. As for myself, I can see that there is plenty of room on the cot. Move over, Rick." Rick settled awkwardly on his side in the four-inch space allotted him and dreamed fitfully of missionaries and the comfortable positions they occupied on their wide, firm beds.

Gitangda Is Great

by Monique Borgerhoff Mulder

> *"Prepare beer 7/14/88. 7/15/88 slaughter sheep and prepare* loghamajeg. *Following Wednesday 7/16/88 set off for Fuweid, returning 7/17/88."*
>
> Village Chairman, Milanda 6/27/88

THE VILLAGE CHAIRMAN put down his pencil stub, removed sunglasses made from the bottoms of two beer bottles, and extended his hand. Our written arrangements were formalized with joint signatures. Honey beer was to be prepared on the 14th of July; a sheep to be slaughtered on the 15th, and the *loghamajeg* skin strips to be cut. On the evening of the 15th we were to arrive at his house, and at dawn on the 16th start our great pilgrimage to the Ngorongoro Highlands with three Datoga elders. Our purpose was to pay homage to the grave of Gitangda, apical ancestor of the Daremejeg clan, spiritual leaders of the Bajuta, a subtribe of the Datoga. Gitangda had been killed in battle by the Maasai at a critical juncture in Datoga history in the middle of the last century—their eviction from the lush, rich, green Ngorongoro Highlands.

The Datoga, Tanzanian pastoralists who herd their cattle on the dry plains in the north of the country, honor their dead in a most elaborate fashion. At a man's death, his family starts to prepare a *bungeid*, the Datoga name for his funereal monument. Over a period of anything from three to eight months a pyramidal structure of stones, sticks, and earth is built over the deceased's body in a corner of his homestead. After the completion of the bungeid construction several hundred relatives, clan members, and neighbors come for dancing and feasting. This very ritualized celebration can last as long as a month and finishes with the eldest sons of each of the deceased's wives climbing on the bungeid and placing grass, honey beer, and tobacco on its cone. After these events the visitors leave and within a few days the family of the deceased abandon the old homestead. With time the eight-foot fence that protected the cattle at night from the attack of lions and leopards sinks to the ground and the flat-roofed huts fall to the ravage of termites, with the result that only a weather-beaten bungeid stands as a beacon to the old man and his ancestry. The bungeid, however, is not forgotten. For many years to come, sons and their descendants, clan members, and even Datoga of different clans and subtribes will travel to this site to pay their respects at the funeral monument of a revered ancestor.

Unknown to the visitors and indeed most Tanzanians is that one of the two most sacred Datoga *bungeidinga* (plural of bungeid) stands on the floor of the Ngorongoro Crater—attesting to the long history of human habitation of an area now renowned primarily for its wildlife. Ironically, traditional herdsmen are now largely prohibited from entering the crater. To learn more about the Datoga's early nineteenth-century occupation of the Ngorongoro Highlands (known to the Datoga as Fuweid), their eviction from this sanctuary at the hands of the Maasai, and to observe the rituals involved in their honoring of an important ancestor, I decided to organize a motorized pilgrimage, starting from the eastern shore of

Lake Eyasi, where I was conducting research on Datoga pastoralism and family life.

As an anthropologist who had only just begun working with the Datoga, I saw a number of potential advantages to this escapade, other than what I would of course learn about Datoga history and culture. Most important, it would gain for me some much needed street credibility. The Datoga, like most livestock herders in East Africa, are justifiably proud of their ability to thrive in an environment in which agriculturalists (of which I am an especially sorry specimen in their eyes—my father owns less than five acres which we don't farm) wilt. We are constantly digging thorns out of our feet and gasping for water at the end of a day's walk. Furthermore, since we are unable to wield a spear, we are derogatorily referred to with a Kiswahili phrase *chakula cha kuki* which means food for the spear, on which we sometimes presumably end up.

Luckily, from the start of my project in 1987, some Datoga families had been kind to me, inviting me to live in their homesteads. They nevertheless took pleasure in criticizing everything I did. My feeble attempts to make a cow, terrified at the sight of a white person, yield milk deserved such ridicule. But I found it harder to take criticism for skills of which I was more proud, such as driving across rocky river beds or taking notes in almost pitch-dark huts, particularly since no Datoga in the community knew how to do either of these things. I had already learned to live with my apparent ineptitude (anthropologists do well to wear thick skin), but I increasingly sensed a need to upgrade my image. By visiting a sacred site deep in the country of the forsworn enemy neighboring tribe, the Maasai, with a group of respected Datoga elders, albeit in the relative safety and comfort of a Land Rover, I might gain just a little respect, especially if we did everything right.

In respect for Tanzanian Democratic Socialism I left the choice of our guides to Simon, a Datoga who as *Mwenye Kiti* ("him having the chair" is the literal translation of the Kiswahili term) has nominal control over some scattered ten-cell units; these are all that is left of the Tanzanian government's attempts to develop socialist awareness at the northern end of the Lake Eyasi basin. Simon had presented himself to me the year before as a leader, albeit with one of the natural weaknesses that accompany leadership in these remote parts: a fondness for beer. This year, on hearing of my interest in visiting Gitangda's bungeid, Simon studiously developed an itinerary. On this (see the epigraph to this story) I shook his hand.

On the afternoon of the 15th, Momoya, Daniela, and I headed up the lakeside towards Simon's home, as instructed. Momoya had been a friend and ally since the day I arrived at Lake Eyasi in 1987; at seventeen years of age (in 1988), he was one of the very few Datoga young men who had some years of primary education (before the school teacher was murdered) and he speaks good Kiswahili, the national language of Tanzania. Daniela was new to the area; that summer she was initiating a research project on the role of women in Datoga society.

Passing the party office in Milanda we were a little surprised to see a group of six Maasai elders under a tree, drinking bottled Safari beer—a rarity in the Eyasi backwaters. Maasai warriors lurked in the bushes beyond. I looked more carefully and spotted Simon amongst the elders, jubilant. He caught sight of the Land Rover and came to welcome us excitedly. He explained that last month some Datoga had killed a number of Maasai at the top of the escarpment near Endulin, and intertribal relations had become even more tense than usual. He had therefore called a meeting of elders. It was clear that, at least today, the ice had melted. A crate of twenty-five bottles, mostly empties, lay in the shade of the tree. Simon and each of the Maasai elders were delivering elaborate rhetoric on the sanctity of *umoja* (the national slogan denoting unity), usually at the same time, and frequently drifting out of Kiswahili back into their mother tongues, Kimaasai and Kidatoga. The designated clerk for the meeting, a man from a different ethnic group altogether, scribbled frantically, looking seriously confused.

I was hungry and tired, having just come back from a week of exhausting demographic survey work at the south end of the lake. I did not relish the prospect of joining in the revelry. Indeed, knowing my fuse would be short, I decided to stay in the car and wait until Simon had finished his meeting. Daniela and Momoya did likewise. Beers (by now hot) were brought for us. Daniela was regaled with offers of marriage by a toothless septuagenarian Maasai who insisted he was called "Land Rover" (which, due to the common interchange of "r" and "l" in this language, sounded more like "randy lover"). Toward evening the interethnic summit was drawn to a close and we accompanied Simon to the house of the old man who was to lead us to Gitangda's bungeid next day. Empties of course had to be returned, so I decanted the last of several Safari lagers into my large water bottle.

Forty minutes later we were at Hirba's home, an idyllic spot at the northern end of the lake, encircled with duomo palms through which the evening breeze gently rustled. The picturesque elder who, in my mind's eye, was to star on the cover of some glossy magazine such as *Natural History* as a representative of the remarkably beautiful Datoga people, strolled out of his house in a pink hat and a very inauthentic garish *shuka* (cloth wrap). His side-kick, Girgis, an uncle half Hirba's age, looked decidedly surly and sported a floral shirt beneath his more traditional black shuka. Simon, staggering unmistakably as he coasted from the Land Rover door, was rebuked by Hirba: Why had he brought us so late? It was now dusk and too dark to slaughter the sheep. The itinerary was already violated; we would not be able to leave at dawn the next day, as custom dictates for pilgrimages to a bungeid. Simon uttered some weak excuse pertaining to the urgency of his meeting with the Maasai, and the matter was dropped. More importantly, the beer was brewing and had to be tested, so we each took a sample, served in a long ivory-colored ox horn. Datoga beer is essentially mead; we found it mixes nicely with Safari lager. At some stage much later in the evening, Daniela, Momoya, and I left, agreeing to return at dawn the next morning to slaughter the sheep so we could still make an early getaway to Fuweid.

At 4:30 A.M. on the dot Momoya woke me, responding to his magical and unfailing internal alarm clock. I proposed that we should leave Daniela to sleep. She had just done two, two-day focal studies, which entailed staying in a woman's company and recording all her work and social activities. To collect these data Daniela moved into her subject's household and slept in her hut, so long as the woman was willing; most found it great fun. Because this was an exceptionally dry July and the women had much work to do drawing water from distant springs and seepages, they would often get up to grind maize and do housework in the small hours of the night, dozing off again before dawn. Daniela had consequently stayed up most of the last four nights recording activities and was exhausted. In addition, she would not be very interested in watching a sheep be suffocated to death, the traditional form of slaughter among the Datoga. I therefore suggested to Momoya that we pick her up when we returned from the north end of the lake, on our way to Ngorongoro. This however was not acceptable. Some ritual had to be performed at the slaughter, and Daniela would not be eligible to pay her respects to Gitangda if she failed to show up for the slaughter.

We crashed through the bush into Hirba's place soon after dawn and were taken into a hut where a sheep was slowly being suffocated to death. As blood began to spout from its nose, the skinning began. Numerous triangular strips were cut from the chest and then slit with a knife; they were then placed as rings over the right-hand middle finger of each of us pilgrims. I got a reasonably fluffy bit on which the blood had already congealed; Daniela's was long and still very bloody. Finally with a gourd of soured milk, a large pot of honey beer, a saucepan of meat, and some tobacco we left.

Within ten minutes of leaving Hirba's, Simon asked me to stop. All three elders climbed out of the back with their pots and paraphernalia and headed into the bush. Momoya explained something about another bungeid. Daniela and I were horrified; how many bungeidinga are there between here and Ngorongoro? It turned out we were stopping for the grave of Hirba and Simon's father, a great-great-great-grandson of Gitangda; this was reasonable, so we relaxed, and enjoyed watching the golden early morning sunlight creep along the rift valley wall. After half an hour or so the men returned, in fine spirits, and we started off on the perilously eroded tracks that skirt the west and southern sides of Oldeani Mountain, finally reaching the main road up into the Ngorongoro Highlands. Signing in at the Conservation Area gate was awkward: my *loghmed* was much less attractive than it had been in the morning. I had had to top up the rear differential oil, and an oily scrap of rather fatty, fresh goat skin left its mark on the Free Permit Entry book.

We wound our way up into the cold, clouded forest of the crater rim, completed some painless formalities at the Ngorongoro Headquarters (our rather unorthodox visit had been cleared in advance with the authorities), and followed the rim road around to the wide deep slopes stretching northwards to the Serengeti plains. Here you feel like you are on top of the world, as you gaze across the endless savannah that melts into the sky. We stopped at the junction with a steep road leading down into the crater. Hirba, Simon, and Girgis peered down one thousand feet below in anticipation and excitement: Our pilgrimage had now begun in earnest.

We reached the crater floor and headed towards the Lerai Forest, for it was on the fringes of the forest that the bungeid was supposed to stand. Hirba had been here before some forty years ago with his father and a group of singing women. Driving toward Lerai, we stopped several times for Hirba to *piga ramani* (fix a map in his head), which entailed his getting bearings from the numerous cattle paths that zigzag down the crater wall. Before reaching the forest he directed me up to Fig Tree campsite where a magnificent old tree shades a lush meadow. Since the Kenya/Tanzania border reopened to tourists in 1984 this has become a very busy spot with permanently pitched tents, toilets, and showers. Hirba clambered out of the back of the Land Rover, looking exceptionally confused by these new amenities, talked a lot more about piga ramani and disappeared, with his cronies and their spears, into the long grass.

It was hot and I was tired from the six-hour drive; I was also rather disappointed at Hirba's lack of conviction as to where exactly the bungeid stood. From everything Momoya and I had heard earlier, it was on the other side of the forest. I felt that we were wasting time at this particular spot, and went to sit in the shade for a long drink of water from my bottle. It was yesterday's beer but it didn't seem to matter. Forty-five minutes later there was a squeal of brakes as a park warden's pick-up spun on two wheels into Fig Tree campsite with three very frightened Datoga crouched in the back. I was bombarded with accusations: "my tourists" were on foot and not in a vehicle, and were therefore flaunting a principal Conservation Authority regulation that is intended to ensure the safety of tourists. For my lack of responsibility I was to forfeit my Land Rover and be immediately driven up to the chief park warden's office on the rim. Our pilgrimage was off.

It was of course true that Hirba and company had been footing it across the bush, but I was at a loss to know how to plead our case. Datoga live in the remotest of bush country, killing elephants, lions, and buffalo with only a spear. These were no blue-rinsed American dowagers venturing out to pat a lion, and they hardly seemed at great personal risk, given their ancestors had survived in this very crater for centuries. Inevitably, of course, I played abjectly apologetic: what a senseless, irresponsible, cruel *mzungu* (white person) I was, sending my servants out at such personal danger to look for a private campsite for me (this was, of course, what the warden thought I was up to); how kind of the warden to retrieve my men, etc. After a good deal of this the warden left, convinced that Europeans were even more stupid than he had previously imagined. We had a good laugh, particularly over Hirba and his crews' smart pretense of failing to understand even a word of Kiswahili, the national language of Tanzania.

After the warden's departure, Hirba, rather shaken by the events of the last hour, confided privately that he was having some problems with his bearings. The shape of the rim and the lines of the cattle paths suggested to him that the bungeid was nearby, but he, Simon, and Girgis had searched the area high and low before they were picked up by the warden, and they could not find it. I drank some more "water" and came to a quick decision; it was already 5 P.M. Rather than search on the three other sides to the forest, we should go straight to the rangers' forest post and ask if any of the rangers knew where the Datoga bungeid was. This was dreadfully humiliating to all concerned, but it seemed the only feasible option if we were to get settled at our shrine before dark.

At the Lerai cabin, adjacent to the ranger station, there were some strange goings-on. First, there was a white woman preparing an enormous fruit salad out of passion fruit, papaw, pineapples, and mangos, apparently for nobody—at least there was no one around. The sight of this enormously cheered Daniela, who had been sunk in gloom and exhaustion since 4:30 A.M.; we hadn't eaten anything fresh for a long time; it was nice just to look at the cut fruit. Next, gathered in the shade behind the cabin we spotted about seven young men. Momoya must, I think, have thought they were off-duty rangers because, pointing towards the forest, he asked if any of them knew where Gitangda's *kaburi* was; kaburi is a pretty reasonable translation of bungeid into Kiswahili (it means grave). At this question, the young men swung into a frantic excited cross-examination of Momoya: "Who? Where is he? Does he live here? Is he still alive? Can we go and see him?" Momoya backed off nervously. I subsequently found out that this was a group of West African students with only a smattering of Kiswahili. They must have been anticipating quite a spectacle, perhaps the remains of a body recently gored by a rhino.

At the ranger station itself we found a very helpful man, who knew exactly what we were looking for and offered to guide us. He happened to be Maasai. He climbed into the Land

Rover, casting withering looks on the four Datoga unable to find their own sacred site. The elders kept their eyes fixed on their spears and the gourds of honey beer. The ranger, a very educated man, explained to me in English that the Datoga would occasionally come down to the bungeid in the crater to sing and make offerings, so he knew the site well.

He directed us straight back along the track to Fig Tree campsite, where Hirba had first alighted. Here we left the car and followed our Maasai guide to the showers, from the showers to the toilets, from around the back of the toilets into the deep grass where people went when they didn't quite make it to the toilets in time, from the deep grass to the rubbish pits, and then finally, from around the back of the campsite dump, picking our way over Kimbo cooking-oil cans, we reached a small heap of stones—Gitangda's bungeid, with its own empty tin of Blue Band margarine rampant. Suddenly there was a flurry of activity. Hirba had all of a sudden got his ramani straight and knew exactly where everything was. In his words, "the toilets and showers disappeared back into the future." He and the other two elders raced to the car and came back with their equipment. Leather sandals were kicked off, beer was gulped and then spat over the stones, same for the milk; tobacco and meat were placed between the stones, and loghamajeg skin strips were attached to the rough face of the small monument. By the time Daniela had collected her camera from the Land Rover it was all over, and anyhow the light was bad and the rubbish heap was most unsightly.

Then there was a wild scamper, gourds and sandals swinging, across the campsite to the huge fig onto which more loghamajeg were pinned and the same victuals offered. This horrified the Americans enjoying their sundowners in the comforting shelter of the great tree; it amazed the safari company cooks who were preparing the tourists' beef bourguignon on a fire set in the roots of the sacred tree. Little did any of them know they were camped at the shrine of Magusachand, father of brave Magena who subsequently saved the Datoga from the continued ravages of the Maasai after the death of Gitangda. Daniela and I struggled unsuccessfully to get pictures of these rituals in the dying light under the vast dark fig canopy, between taking off our own shoes and spitting beer and milk in the requisite spots. The elders, at least, were content with their work and settled down to the gourds of honey beer. I drove our Maasai guide back to his post and returned to Fig Tree campsite to face the interethnic music.

The tourists' tents were already pitched and their food was being cooked as they tranquilly gazed on the Ngorongoro Crater, "a spectacle of a lifetime," in the golden evening light. We rather upset the tone of things. Up went my old army tent, and outside it sat four Datoga wrapped in their dirty blankets, with two equally dirty Englishwomen, all drinking large quantities of honey beer out of ox horns and getting progressively more raucous. At some stage during the evening—about when the safari group, largely comprised of elderly ladies, were about to be served their dinner—the elders demanded their food. Under the invaluable instruction of Momoya, we chopped up the sheep (not a young one) with an ax, and prepared the fire and *ugali* (a staple maize porridge). Momoya exhibited his limitless skills in cross-cultural dialogue by getting onions and salt with which to increase the palatability of our efforts from the Mbulu cooks of the safari company. When the food was ready we brought water for the men to wash their hands and served our meal. After much yarn-telling deep into the night about who Gitangda, Magusachand, and Magena were, what they did and how they died, the men retired into my tent and Daniela and I snuggled up with the toolbox and mutton bones in the back of the Land Rover.

The next morning it was exceptionally cold and the old men, taking their cue from the safari campers, demanded early morning tea, and subsequently more ugali and meat. Daniela and I did the honors again and then, as we were clearing up this feast, the old men nipped off back to Gitangda's bungeid to make their parting offerings of beer, etc., each taking home a small handful of soil from the grave as a souvenir. Somehow between scrubbing the pans and taking down the tent, we again managed to miss valuable anthropological data and photo footage. Simon also took the Blue Band can (it would be useful at home, he said), and the other two raided the rubbish pit and came back with bottles, cans, and half a pot of pili-pili sauce (a wonderful hot relish that turns any meal into a feast). This was a bonanza—never had Gitangda so much to offer!

The back of the car was by now a stinking pit of slopped honey beer and sour milk, dirty calabashes and rubbish, but everyone was happy. We had apparently done everything just right. There was certainly no mention ever again of the help we received from the Maasai ranger. After all, had we not reached the spot, consumed sufficient quantities of beer and soured milk, made the appropriate offerings to Gitangda and Magena, decorated the graves with loghamajeg, and recounted heroic tales of our ancestors into the night? Was this not what a successful pilgrimage required? There was much serious discussion on this question, and unanimously it was agreed that we had succeeded on every count. It seemed that, with the Datoga at least, conducting rituals is a lot less demanding than one might suspect from reading anthropological texts, where ceremonies are laid out as a list of prescriptions, not the (let's be honest now) bungled amalgam of memory-dredging, improvisation, and chance that I was witnessing. I was learning a lot about what ritual really entails, with its intriguing blend of solemnity and fun.

As our discussion drew to a close, it emerged there was another requirement of pilgrims, and this was to return home immediately when the business was successfully completed, to ease the fears of the loved ones left behind; Maasailand after all is not a safe place for a Datoga. On this particular occasion, however, Simon proposed that a little "game drive" would not constitute a serious infringement. A quick loop around the Gorrigor Lake, which the Datoga knew to be the core home range of a large pride of lions, was agreed upon by all. Daniela was thrilled. She had never been in the Ngorongoro Crater before, and until now it looked like all she might see of this jewel, aptly dubbed "the eighth wonder of the world," would be its rubbish dump and toilets.

Momoya too was delighted. He has a limitless ability to adapt to all cultural contexts. We could hang out happily together, whether as free guests entertaining tourists at safari lodges, on month-long stints in the bush with the Datoga, paying courtesy calls on party officials in Arusha, or at the post-seminar dinner parties of Serengeti Wildlife Institute scientists. Momoya is apparently at home everywhere, initiating conversation and asking stimulating and provocative questions of everyone he comes across, so long of course as the conversation stays in Kiswahili. His infinite social skills never failed to amaze me.

But he follows one rule that helps a lot: a variant of "When in Rome. . . ." And he had followed it that morning. Knowing that white people love to have showers at every opportunity, he reckoned that as a camper at Fig Tree campsite he should take advantage of the facilities, and accordingly headed off to the little shower hut in front of Gitangda's bungeid. This is indeed what the American and European visitors to the Crater do, but in the heat of the day, not at 6:00 A.M. when the Crater is shrouded in freezing fog and bitter chill. There is of course no hot water. Without a towel he had dried himself with the thin cotton shuka that he subsequently wore and three hours later, despite the tea and mutton, he still looked blue in the face. The idea of a game drive, while the sun gradually rose in the sky, pierced through the early morning mist, and warmed the metal flanks of the Land Rover, strongly appealed to Momoya.

We didn't have to go very far before everyone was satisfied. The first scan through binoculars over Gorrigor Lake revealed a dramatic sight; a lion and lioness consuming a hippo carcass in the reeds. Daniela and I sat up front with the Datoga in the back. We could approach quite closely and our passengers turned frantic with excitement. A Datoga man can earn up to fifty cows and a reputation for life from spearing a lion, if he has witnesses. But lions, like all wild animals, are of course protected in the Ngorongoro Conservation Area. The threat of ever-tightening knuckles on the spears did not, however, unnerve me; I had lain all four spears safely under the spare tire, in anticipation of just such an encounter. We, and finally the lions, left safely, with no Conservation Authority regulation violated.

More delighted than ever, we decided to call our game drive to a close and headed for the "up" road, slowly climbing the tortuous crater wall and watching Magusachand's tree get smaller and smaller beneath us. Back at Hirba's the following evening (we had to delay our return to replace a broken spring leaf in the local town), our arrival stimulated considerable sprinting through the bush. The neighboring women who were supposed to welcome us pilgrims with singing and dancing had given up and gone home when we failed to arrive the previous night. However the car's engine, which could be heard for miles over the Eyasi Basin, drew them again to their posts. It was a pretty sight, arriving at sunset on the side of the lake, to see the track to Hirba's gate lined with women in their beaded leather finery, singing songs about the bravery of Gitangda and Magena in their battles with the Maasai. We relaxed in the cool evening air and were served more honey beer which we drank with the men of the local community.

To some extent we were heroes ourselves, having done everything as custom dictates. The game drive, the Maasai ranger, the drunkenness the night before we left, these things were all forgotten. Momoya had lost his loghmed; he probably left it in the shower (this he whispered to me as we reached the crater rim—I refused to go back down for it), but luckily no one had noticed, for it is a serious violation of custom not to wear a loghmed for three days following the pilgrimage; he was now drinking beer with his left hand. In everyones' eyes, and even our own, we would bring a measure of peace and prosperity to the Datoga of Eyasi, on behalf of Gitangda and Magena.

Finally the singing was over, and we were left to drink deep into the night. The women from neighboring households went home—one with a Blue Band can, another with a half-full jar of pili-pili.

Gitangda is great, or so the legend goes. Certainly the woman with the pili-pili believed it.

Bush-League Medicine

by Kate Kopischke

WAILING AND MUFFLED confusion woke us abruptly the morning Doris' baby died. The boy stopped breathing only a few hours earlier, and we learned the grieving parents were off to the forest to bury him.

Hillard Kaplan's face fell into his hands. He was with the child the afternoon before, had listened through a stethoscope to his tiny, wheezing lungs, and treated him with oral antibiotics. Satisfied that the penicillin would improve the baby's condition, he left, intending to return the next morning for a checkup.

"I didn't treat him aggressively," Hillard still recalls. "I ought to have gone back in the evening, given him an injection, and stayed with him."

The two of us stared, remorseful, into our cold hearth as darkness withdrew from the rain forest. Moments later came another unsmiling visitor. It was Ernesto, informing with urgency that his two-year-old, Timoteo, was struggling to breathe. Hillard leapt into motion, grabbed the stethoscope and kit of injectibles, and trailed after the worried father.

Thus began the 1988 pneumonia epidemic that disrupted life in Yomiwato for five days and nights. Machiguenga villagers came that morning to our hut, one after another, making known their relatives' conditions and requesting immediate doctoring. At once we tabled our data collection agenda, charted the previous year in the anthropology department at the University of New Mexico. Villagers, too, put their regular business on hold; few hunted, fished, collected, or gardened. Sickness enveloped Yomiwato.

When Hillard returned from his house call to Ernesto's, we commenced a systematic survey of the village, administering suitable treatments for pregnant women, small children, babies, the elderly. Day and night we came to the patients' huts, urging full rounds of antibiotics even for those who insisted they'd recovered.

Ernesto, for one, saw Timoteo breathing easily and declared the pills no longer necessary. Fewer than three days into the child's treatment, he thanked us for curing the boy and asked that our visits stop.

But fear that his symptoms would return, that Timoteo would be next for a forest burial, pressed us to continue. Outbursts occurred at the very sight of Hillard, who twice had terrorized the boy with needle pricks to his tiny bottom. Every day we stressed the importance of a proper course of antibiotics to Ernesto, who, with the same regularity, would slip the tablet under his son's sleeping mat and assure us the child would swallow it when his tantrum passed. But Hillard would only add to the shrieking child's anguish; he'd hold the boy steady, retrieve the hidden pill, and coax it to the back of his angry mouth.

We suppose Timoteo will forever recall the white anthropologists who came to his village with the intention of torturing him. And his parents, we are sure, still find humor in their son's fear and loathing. To them, the distress we engendered in Timoteo yielded an infallible formula for convincing their naughty child to behave.

"If you don't straighten up," they would threaten, "we will fetch *Dr. Julio.*" Horror would overtake Timoteo and promptly his disposition would sweeten.

Little Timoteo was not the only villager to brush so nearly with death. Old man Pedro, already well beyond his healthiest years, was medicated back to life by way of regular house calls and a week-long series of injections and oral treatments administered by Hillard. It was no easy undertaking; Pedro's hut was forty minutes from the village proper, reached by a barely discernible footpath thick with lianas, fallen trees, and tropical undergrowth that teemed with unfriendly wildlife.

One evening, following a villager's report that Pedro was gasping his last breath, Hillard trekked in a midnight downpour—medical equipment in hand—to tend to the ailing elder. Such a journey to rescue a man of Pedro's advanced age was, to the Machiguenga, outlandish and amusing. Later, when the epidemic had passed and Hillard divulged the fear that gripped him during his late-night excursion, the saga roused incessant laughter.

The high-point of Pedro's story, for the men who routinely told it, was not Hillard's angst in the dark jungle but his mimicry of Pedro's condition that night. Huddled in a fetal position near a weak fire, the old man breathed loudly and with great difficulty, gasping desperately for any iota of air his obstructed lungs would absorb. Hillard injected him without delay and for the rest of the night the two sat in silence while Pedro pulled slowly, miserably for air.

At first light Hillard administered a second injection and returned through the soggy forest to the village proper. There he updated curious villagers on Pedro's state. The old man was a little better, he reported, but earlier his suffering had gone something like this: "*heeeehuh, heeeehuh,*" and Hillard pulled and heaved to demonstrate Pedro's misery.

Tears streamed from the eyes of the guffawing men, who would implore Hillard to repeat the impersonation once more. Then each man in the gathering took his own turn at imitating Pedro, and more laughter would soar through the quiet village.

Throughout our year-long stay in Yomiwato, remotest of the contacted Machiguenga settlements in Peru's Manu River basin, a string of other illnesses jarred this tropical world. I'd come to the region to work with Hillard, whom I later married, and two other anthropologists—Mike Alvard and Teslin Philips. Our team was there to investigate how parents divide their time among the assortment of inevitable child-rearing tasks, and how a parent's involvement affects a child's growth, development, health, and well-being.

Mike and Teslin, working in a Piro village some four days downriver from Yomiwato, endured equally as puzzling medical episodes—memories of which bring heartache, laughter, and endless distress over the state of healthcare for those remote Indian populations.

Of the 100 Machiguenga living in Yomiwato in 1988 and 1989, more than half had never left the Manu's isolated headwaters. Infant and child mortality was nearly forty percent, and diseases like the pneumonia outbreak—brought by outsiders and by villagers returning from travels to economic hubs downriver—had become commonplace. Attempts to acquire medical training, vaccines, and remedies had failed, and in 1988 the Machiguenga were at their wits' end.

In exchange for allowing us a home and field site in Yomiwato, the Machiguenga had one request: medical assistance. Forest remedies, they complained, were not keeping their children alive, nor were they effective against viral assaults like pneumonia and other calamitous illnesses that plagued the region with greater frequency.

To us, the request was more than a trade for services. It was a matter of urgency that the Machiguenga be granted the same healthcare privileges as other Peruvians, and we willingly agreed to help bring a change. Indeed, our decision met great criticism from foes of Western medicine—purists who charged us with corrupting native ways and with callous disregard for the healing powers of the rain forest's extraordinary flora. Critics sought to revoke our research permits, and at regular intervals we learned of rumor campaigns designed, undoubtedly, to instill distrust among the Machiguenga themselves.

But we remained steadfast in our commitment, doctoring as best we could and campaigning for an official program of public health. To the Machiguenga, the concerted efforts to deny medicine, and to force Manu's remote communities to live the romantic ways of the past, was paternalistic and ignored their right to self-determination.

Some years earlier, a baby had died in Hillard's arms in an isolated Paraguayan settlement. It was then he vowed never to venture to any field site unprepared for such tragedies, and decided to learn the basics of tropical health care. The training became a requirement for students and field assistants.

Before and after our arrival in Yomiwato, we pored over notes from lessons in primary health care from Ben Daitz, a friend and professor of medicine in New Mexico who himself had visited and delivered health care to the Machiguenga. With regularity we consulted Manson's *Tropical Disease Manual*, an assortment of medical dictionaries, and the latest edition of *Donde no hay Doctor.*

After doctoring our way through the Yomiwato epidemic, managing somehow to avoid more death among the thirty or so gravely ill villagers, we felt sufficiently equipped to handle any medical challenges that lay ahead.

Our inventory, planned with Ben's careful appraisal, included a full range of antibiotics, aspirin, and antihistamines. There were bandages, tongue depressors, sutures, syringes, thermometers; a case of rehydration formula and generous supplies of the favored pink Pepto Bismol pills; anti-fungals and anti-malarials; two stun guns; a box of pregnancy tests; and a smattering of Valium and other sedatives for the unknown odd exotics.

But there was nothing in the supply bag to calm the illness that overcame Hillard for nearly the whole of March. We talked by radio to doctors in Lima and the United States, explained his symptoms of fever and chills, unrelenting head and back pain. They urged him to leave immediately for a hospital in Cuzco or Lima—until he described to them the wearisome journey out. After a long boat ride to the disagreeable port town of Shintuya, through the thickest of the Manu's mosquitoes, sandflies, and impossible log jams, one is still several days from civilization. To reach Cuzco, we would set forth on a frightful truck ride—crowded, comfortless, jarring—through the rugged Andes. On a good journey the trip lasts some thirty-six hours. But bad journeys are typical.

Hillard would not attempt the excursion, opting instead to wait out his illness in the comforts of Yomiwato. Seeking air and a respite from his sweat-saturated foam mattress, he would wander from household to household collecting data or endeavoring to complete a demographic blueprint of the region. Other days he tried reading or writing in the shade of a palm thatch, but he worked in utter confusion. Once during his delirium he pondered how we might transport his corpse downriver, out to Lima, and back to New Mexico. In the likely event of his death, he announced, I was to cut his body in two, place each half in a river bag, and deliver him thusly to his parents.

His pain was spectacular. The headaches were all-consuming, enough to persuade him that permanent brain damage had transpired. Realizing the ineffectiveness of Dr. Julio's own medical kit, villagers grew worried and began debating appropriate local prescriptions. Miguel and Ana, Yomiwato's mission-trained school teachers—themselves Machiguenga—arrived with two of their closest attendants to administer a local remedy.

A billion stars decorated the sky over Yomiwato when the operation began. Inside our hut, in the dull light of a twelve-volt lantern, Ana began issuing orders: "Cut these leaves," she told her assistant, Victoria, while handing her a stack of oblong, flat leaves from the nopal cactus. "Katrina," she commanded me, "I need ten sheets of paper from your notebook, and I'll need a fork."

Victoria and I worked obediently, she slicing down the leaves' thin centers and I curiously freeing paper from spiral. Hillard's pleas for me to show the medical team out, to convince them the fever was gone, were in vain. I was far too desperate to abort this intriguing initiative, particularly since there appeared little chance that the slime they planned to administer would exacerbate his condition.

Following Ana's next directive, I laid out the ten sheets of notebook paper, one next to the other across the wide board that doubled as desk and kitchen table. Close behind, Ana stabbed at the paper with the fork, poking dozens of holes uniformly into each sheet. Next she took the open-faced halves of sliced nopal leaves and smeared their oozing contents onto the punctured notebook paper, half a leaf's slime to each sheet.

"Katrina," Ana said sternly, "you must watch carefully and repeat this process twice before morning. If you do not, Dr. Julio will not recover."

She then climbed cautiously into the tent where Hillard lay, and behind her an assembly line formed. Twelve-year-old Graciela, youngest of the attending aides, handed two of the slime-laden sheets to Miguel, who passed them to Victoria who passed them to me. I examined them carefully under the solar-powered, fluorescent bulb that lit the operating tent; they were heavy, cold, and so bright green they nearly brought cheer to the solemn procedure.

I handed the first two applications to Ana, again ignoring Hillard's petition to be let alone. On each side of his head she gently pasted an ointment-covered notebook sheet, followed by two more affixed symmetrically to his underarms, two to his stomach, and two to his upper thighs. Hillard squirmed and complained, his cultural sensitivity wearing thin. I made the case to him that the Machiguenga were as desperate as we were; that it would cost little to abide by the remedy and complete a full treatment.

After a lengthy discussion in Machiguenga with her aides, Ana urged me again—in simpler Spanish—to carry out the process twice before sunrise.

"*Sí*," I promised halfheartedly, knowing Hillard would wholly refuse further treatment. The team stood to take its leave and as the last footstep crossed the hut's threshold, Hillard—slime covered and miserable—tore frantically at the sheets, removing them from his head, torso, and legs, and crumpling them into a sticky pile beside the mattress. He peeled off his shorts, now covered with the drying goo, and ripped away the bedsheet, onto which significant globules of the slime had dripped. In an instant he was standing beneath the solar shower rinsing himself clean of the emollient, declaring in no uncertain terms that the experiment was over.

On rounds to our hut the following morning, Ana and her team found the patient recumbent, as usual, in the same feverish sweat they left him in the night before. Nearby lay a stack of the unsliced nopal leaves and the crumpled notebook sheets from Ana's initial application—clear evidence that treatment had not been carried out as directed.

Ana's declaration then that Hillard would not recover reduced me to deeper gloom. I thought at first she was right, that unless we evacuated him he would perish here in the rain-soaked Peruvian outback. Until his eventual recovery, from what we now believe was meningitis, Ana would scold me for my failure to follow her instructions. When the headaches and fevers finally did subside, she modified her reprimand: They'd known all along Hillard would come around, but our negligence in administering the nopal slime had greatly prolonged his illness.

It was not the first time our bush-league medical practice failed to produce a cure. At Christmas the doctoring we had attempted was decidedly ineffectual following a horrific snakebite episode—another occasion that paralyzed Yomiwato for days.

By the community's own diagnosis, Teresa—thirtysomething and eight months' pregnant—ought to have died from the attack of the seven-foot fer-de-lance, whose fangs pierced the flesh just above her right temple. The puncture resulted in six days of extreme swelling, nausea, and agony. When we first saw Teresa after the bite, we believed we could help her with some remedy from our kit. But in a few days' time, we joined the community in believing she would die.

It happened during one of those sultry, wet-season days, not peculiar to Yomiwato. Teresa rose before dawn, prepared a meal of banana pap for her husband and three small children, and at first light set out to collect the bark she and her husband would spin into twine for tote bags and hunting bows.

The family walked more than two hours before spotting the bark source. As Teresa reached down to begin tugging at the tree's surface, the fer-de-lance struck, stabbing her neatly with both teeth.

Carlos hurried his wife and children back across the thorny, winding trail that had delivered them to the evil viper. In the two hours it took to reach Yomiwato, its poison progressed through Teresa's body.

On the porch at Miguel and Ana's hut, caregivers shaved away a section of her long, thick hair and placed on the wound the village's mysterious black stone, a gift of some years back from a Frenchman called Dedier. According to local lore, Dedier had distributed these smooth healing stones to communities throughout Amazonia in an apparent effort to keep natives in touch with the more natural and mystical practice of traditional medicine. The rock contains unknown properties that cause it to stick to swollen skin, and it is believed to be capable of sucking out whatever poisons or infections permeate a patient's body. We once heard the stone was mined in Switzerland. That it was not native to the jungles of South America seemed to bother no one, and in Yomiwato, belief in its power was staunch.

After affixing the stone to Teresa's wound, Miguel and Ana sent messengers to our hut with the tragic news. We sped to their house with supplies from our medical store: stun gun, venom extractor, bandages, alcohol.

Unaware then of the stone's huge significance, I removed it hurriedly from Teresa's head, set it on the floor and applied the venom extractor to the area around her wound. Twice I tried to remove fluid, but none remained at the surface. We squeezed the wound, applied alcohol and squeezed again, but to no avail. After weighing the potential costs of zapping her with the stun gun (a shock treatment that many experts believe neutralizes the venom and prevents it from spreading), we decided against it, fearing a shock to the head of a pregnant woman would do more harm than good.

As we stood shifting back and forth, looking to one another for new ideas, a small group gathered around the little black stone, plucked so hastily from the victim. Whispering rose to excited argument and at once I knew I had erred. It was Victoria who finally approached me.

"Katrina," she said timidly, "we need some milk."

"Milk?" I queried. "Whatever for?"

I could not interpret her complicated answer, yet the urgency in her voice was plain. In five minutes I'd returned to the porch with a single can of condensed milk. Victoria snatched it at once, plunged into the metal with a rusty knife and poured a small amount into a calabash bowl. She placed the stone gently in the bowl and swished the milk round and round, a dozen eyes peering steadily from above.

More villagers gathered to watch Teresa's swelling intensify. Relatives remained positioned around the soaking stone, whispering, waiting for the perfect moment to pull it from its milk bath and return it to the victim's wound. Now we began to understand the huge importance of the rock, called upon only in the most dire situations.

Teresa's case was indeed dire. Hers was as bad a bite as most villagers could remember. Many in Yomiwato had lost loved ones to snakebites, and they doubted Teresa would survive this one. Yet the stone seemed to offer a glimmer of hope. Victoria removed it finally from the calabash bowl, dried it on her skirt and reaffixed it to Teresa's throbbing, swollen head.

Helplessness overcame us. In despair we began second-guessing our decision not to use the stun gun, and Hillard dashed to the radio to attempt contact with the Center for Disease Control or any other help agency that might advise us. Courtesy of the halo frequency, a network that supports the fieldwork of missionaries throughout the world, we were patched to the CDC that afternoon. Experts there, as well as other ham-radio doctors who joined in the effort, advised against use of the gun.

Whatever treatments were recommended we had already tried, and Teresa's condition grew worse. Realizing that the Western bag of pills and potions contained nothing for the patient, the Machiguenga turned to Chaco, a wise and respected elder, to serve as Teresa's healer.

By the following morning her face had bloated so completely that her eyes swelled altogether shut. The decorative bands around her wrists and upper arms became too tight to tolerate, so her sisters cut them away. Throughout the day and night family members and friends sat quietly by, watching Teresa suffer. Every few hours Chaco prepared cups of hot *piri piri,* a grass-like plant used throughout the region to cure an array of tropical ills.

To an outsider the treatment seemed cruel. Each time Teresa swallowed a portion of the piri piri, she turned to vomit in a small hole dug in the dirt beside her. The idea, we surmised, was to rid her of the evil fluid infiltrating her system. But to us the wretching and nausea seemed only to heighten her misery.

It was a painful time for everyone. Chaco, who had long served as Yomiwato's medicine man, claimed to have lost all power to heal. I looked at the sad wrinkles in his face and saw his desperate longing to help his niece, to regain—just this once—the power he held as a younger man.

Teresa's younger brother David, who joined us for a meal of fry bread at our hearth three days after the bite, was accused of prolonging Teresa's suffering by eating margarine. Ingesting greasy foods during a family member's illness is a serious taboo among the Machiguenga, and when we discovered this I assured Chaco that David's piece had in fact been served without margarine.

Desperate for answers, some family members and other caretakers began blaming Pablo, another of the victim's younger brothers who was away on a fishing trip and unaware of his sister's condition.

"It must be Pablo who is eating sungaro (an oily whitefish)," David charged. "It is his fault our sister is not recovering."

On the afternoon of Christmas Eve, the mood in Yomiwato reached its grimmest. Since the day of the snakebite, none of Teresa's relatives had hunted or collected food; bananas and manioc and an occasional pot of noodles or rice barely kept the extended family nourished. Fatigued but restless, they continued their watch over Teresa. Late in the day, a family of black rainclouds floated in to form a grim canopy over the village.

I sat alone at our hearth, uneasy in the calm air, preparing dough for two loaves of cinnamon-raisin bread. Sonia and Ermalinda, two delightful and curious young girls from the hut next door, spotted me and came to watch. To them, bread making was as foreign as Christmas and New Year's, and they studied the process intently. They marveled at the tin cracker box we used as an oven, though to them the biggest thrill was devouring freshly baked slices.

The girls hadn't seen the cinnamon-raisin variety before. It was a treat for the following morning, I told them, because where we come from December 25th is a special holiday for feasting and celebrating. I finished the kneading and placed the loaves aside to begin the long rising process.

Just then, as an enormous cloud swallowed the sky above Yomiwato, a procession led by Carlos and Teresa moved in single file past our open-air hearth. The children followed, then her brothers, sisters, and a long line of relatives and friends. Chaco was last.

"The storm is almost here," he told us. "We are moving the bitten woman to a house with a better roof, where she will be protected." Then, as if to update a colleague on a patient's condition, he told Hillard: "Tonight she will die."

With sunken hearts we stood silently, stared at the dirt until Chaco walked slowly on. Pacing in the clearing in front of our hearth, we wondered how we might have made a difference. Should we have worked harder by radio to airlift her to Cuzco or Lima? Attempted the boat ride to Shintuya?

We rounded up some candles, a small pillow, and a thin camping mat and walked next door to offer them. Perhaps these would make Teresa more comfortable, we suggested feebly. Then we sat, joining the large group that surrounded Teresa, whose head continued to throb, eyes remained buried in the folds of her puffy, translucent skin.

Without warning the wind rose into a violent rage and a torrential rain began battering the small village. Rooftops blew off surrounding huts; trees bent sideways until they snapped, crashing into the churning river. A spectacular clap of thunder flattened the rising cinnamon loaves.

We never slept that night. As the minutes ticked gradually by, we sipped from a small bottle of whiskey, one we had hoped to share under happier circumstances. We listened hour after hour as the rain pelted our thatched, leaking roof, believing that at any moment Teresa would gasp her last breath.

Dawn finally crept through the thick, drizzling jungle. From next door a loud commotion arose and we rushed to investigate. To our astonishment, Teresa sat upright, eyes open for the first time in six days. Her children played in the clearing around her, tossing the shiny black stone that had dropped from her head only moments earlier.

"The woman will live," Chaco told Hillard as matter-of-factly as before.

"She will live," Sonia repeated to me. "Now she is fine."

My throat clogged. Sonia gripped my arm and held tight, even as I scooped Teresa's two-year-old into my arms.

"Unbelievable!" I said to Sonia and the little one. "Do you know this is Christmas morning?"

Sonia grinned oddly, looking at me with confused eyes. The little one squirmed to the ground and ran off.

"It's Christmas, *Navidad,*" I repeated to Sonia, who spoke no Spanish.

"Katrina," she asked, smiling meekly and still holding my arm, "what is *Navidad?*"

We still laugh at the notion of trying to explain the virgin birth in Machiguenga. Soon after the holiday, Teresa delivered a seven-pound baby boy and life in Yomiwato, for a time, returned to normal.

Section V: Research Communities

In the Forest Without a Dog

by Andrew Grieser Johns

THE MOTOR CANOE grounded on the sandbank with a jar that pitched the bowman overboard into the shallow water.

"Where are we?" I asked.

"Just some stream or creek or something," said my assistant helpfully. "There aren't any maps."

There was a second splash as the fellow manning the high-powered outboard engine also fell overboard, having been seriously depleting a private supply of *cachaca* (locally brewed cane spirit) on the long journey upriver.

"If we lose him we're really stuck," I observed. He was the only person who could work the engine and I couldn't remember having seen any paddles in the canoe.

"We," intoned one of my fellow scientists seriously, "are in the forest without a dog."

"Portuguese idiom," explained my assistant. "And don't put your foot into the water, there'll be piranhas."

"They didn't touch the two who already fell in."

"They're from Paraná."

"So what do we do now?"

"Try getting out of the boat for a start. And tread carefully. There'll be stingrays."

"So how do we know this is the right place?"

"This is where the boatman brought us. So it must be right."

My assistant leapt athletically over the side of the canoe and disappeared to his waist in an underwater pot-hole that might have been excavated by an electric eel. If it was, its owner either wasn't at home or figured that cariocas were already sufficiently hyperactive. He gained the bank undamaged.

I followed rather carefully.

Where we were was somewhere off the Tocantins River in eastern Brazilian Amazonia. The Tocantins is a big river. The flow can be 89,000 cubic yards per second at flood levels. Its annual outflow is greater than that of the Mississippi: half as much as the Zaire River. It is the most easterly of the major tributaries that flow north from the central Brazilian shield and eventually merge to make up the Amazon River itself. It is full of bizarre and little-known creatures, many of which bite, and meanders through one of the largest remaining areas of rain forest in eastern Amazonia.

The reason for our presence was an expedition to investigate the ecological impact of a major Amazonian hydroelectric project. It was March 1984. An enormous dam was being built across the Tocantins River at Tucuruí. Rather late in the day, the hydroelectric company had realized that some assessment was required of the impact of the sudden appearance of a 1,000-square-mile lake in what was previously undisturbed forest. A group of Brazilian and overseas scientists had been contracted to undertake baseline surveys of the animals inhabiting the forest in the lake region.

I had been conducting surveys of various types of animals in different parts of the world for some years and the technical aspects of the current job didn't bother me. I did have some worries about the time allocated to cover the land area involved. Generally, however, I counted myself fortunate to be in such an exotic and exciting place.

I had been telling myself this quite a lot during the three-day delay in Tucuruí while we were waiting for transport to materialize. Tucuruí was a typical ramshackle Amazonian boomtown, grown up over a few years to serve the needs, mainly liquid and/or biological, of the construction workers. Apart from such highspots as The Cockroach Bar/Restaurant on the main street, where cold beer was to be had, generally accompanied by the unwelcome attentions of the resident transvestites in which this particular place seemed to specialize, the town had little to offer. All such towns seem to have establishments called The Cockroach, or even The Giant Cockroach. I was told it merely signifies a good place to be: Brazilian cockroaches are a bit picky.

When we finally embarked on the first trip upriver the feeling of general excitement returned. The mud-laden waters of the river were at flood levels. The river was several miles wide and fast-flowing, dotted with clinging clumps of vegetation that marked submerged sandbars. Dead trees and other debris swirled past, smaller pieces hitting the boat with dull thuds before vanishing in our wake. Waterbirds abounded: large-billed terns with Concorde profiles dived for small fry in the shallows, herons and egrets flapped into trees at the river margins, ospreys wheeled overhead. The Amazonian touch was provided by occasional flights of macaws and parrots that passed overhead calling raucously. Leaving our companion boats straggling in line behind, we lost ourselves in the muddy vastness that is the Amazon basin.

Back to reality, I trudged up the sandy beach, already slapping vigorously at the biting sandflies. The boatman, shouting down from a vantage point, informed us that we were in the correct place and could start unloading. Other motor canoes began appearing from downriver. More victims for the sandflies, the head-shrinking Indians, or whatever else lurked in the dense forest I could vaguely see through the haze of mosquitoes hovering over the neighborhood swamp.

As the canoes pulled up on the beach, researchers from all over Brazil began to mill around aimlessly, giving the place the aura of a scientific congress. The more practical laborers began putting up what was to be the first of several tented encampments at various points down the 100-mile stretch of the Tocantins that would disappear within the lake. A completed encampment consisted of two or three huge tarpaulins erected over sturdy poles and covering a framework for slinging hammocks. That was the sleeping area. Several smaller shelters acted as laboratory or work areas and as the kitchen, although these were often indistinguishable. The kitchen, which was generally erected in the best vantage point, inevitably became the acknowledged social area, and the site of some fascinating gastronomic experimentation.

We, the assembled company, represented the first phase of the Tucuruí Wildlife Project: the research operation. Our role was to document the numbers of animals within the rain forest that would end up beneath the lake. As the lake flooded, we pointy-headed scientists would be replaced by the much more macho animal grabbers who would be implementing the animal rescue phase: "Operation Curupira." (The rescue operation was named, incidentally, after a mythical, pestilential little beast with its feet on backwards, whose ostensible role is that of spirit guardian of forest animals but which, as any hunter will tell you, is all too easily seduced away from its purpose by strong drink.)

What made the preliminary research operation attractive at all, to most people, was the provision of a substantial per diem for participants. Researchers fought for places. Encampments swarmed with entomologists; writhed with herpetologists; were plagued with epidemiologists. I and my assistant, who were interested in live animals *in situ,* rather than specimens or component parts thereof in petri dishes, were in something of a minority. The general feeling was that since the animals in the lake area were doomed anyway, it was a chance to forget whatever moral scruples you might ever have had and push forward in the name of Science. The sight of entomologists setting off with chainsaws to collect canopy insects might have seemed a little extreme, but it undoubtedly saved a lot of time and trouble climbing trees. I could see the reasoning, but unlike most Brazilian researchers I was not an aspiring amateur hunter. My assistant and I were in the forest for much of the day and occasionally part of the night too, and saw a great many animals from snakes to jungle cats. But I had no particular wish to grab, trap, or shoot them, or to allow my assistant to do so. Amongst the cytogeneticists, my name was mud.

There is a well-known saying that perhaps originated among the Somalis of the horn of Africa. It runs something like "Those who know it do it; those who know a little, teach it; those who know nothing, direct it." Within an hour of our arrival, several distinguished scientists were arguing furiously with the laborers over the correct way to erect the kitchen. Leaving them to it, I coaxed my assistant away from the array of shotguns, which he was eyeing appreciatively, and we started walking around the periphery of the campsite to take stock of our surroundings. I did not like the look of the swamp. It seemed that we were setting ourselves up on an island isolated from the terra firma forest. No one else seemed concerned. A couple of herpetologists wandered along the edge of the swamp in the other direction looking for snakes that might already be moving away from the site of disturbance. In the shade of a clump of trees, the entomologists were settling down for an afternoon nap. The cook had already borrowed a shotgun and was busy shooting at hawks that were sitting watching the proceedings. My assistant and I set off to try and find a way through the swamp. It had alligators in it.

Some time later, having got lost and been forced to splash around in the dark for an hour or so, we emerged again, relatively unscathed. The company was assembled in the newly-erected kitchen eating dinner. The cook had organized his domain and produced something quite pleasant from meat brought upriver. There wouldn't be much more of that, so we made the most of it. What we were going to eat in the meantime, I had no idea. I supposed that the cook would think of something.

"*Pirarara*," said the ichthyologist, in answer to a somewhat tentative enquiry. "It's a kind of carnivorous catfish. They grow really big and there are stories of people being dragged down into the depths by them. Could happen: look at those teeth. This one is actually quite interesting. It might even be a new sub-species. Have some."

He continued eating it reflectively.

It was very good. I had seen a lot worse lately. I even felt friendly towards the cook. It was a relief generally to get back to camp: be it ever so humble.

Together with my assistant I had spent the previous two weeks at a hunting station some way upriver, where I had been looking for some of the rare monkeys that were to be found in the region. The menu there had consisted of boiled tortoise twice a day. The giant forest tortoises were anything up to about thirty-inches long and had a habit of sunning themselves in the middle of forest trails. With my eyes on the canopy, I was forever falling over them. It wouldn't immediately have occurred to me to eat them, although I understand that there are Indian tribes that eat very little else and thrive on it. There is nothing quite like getting up at dawn on a chill, damp morning and facing cold boiled tortoise for breakfast.

One of the herpetologists wandered in with a small, diamond-patterned snake, which he casually dropped into a bucket by my feet.

"The Indians call this one a 'tenstep'," he remarked informatively. "They say that if it bites you, you can generally walk about ten steps before dropping dead. Tell me if it moves."

He helped himself to some carnivorous catfish.

I put my feet up on the bench.

The ornithologist, who had arrived on the latest boat, dropped her plate and rushed outside as screeching from the trees indicated that the golden-winged parakeets were back. These small, green, yellow-winged parrots were generally believed to be endangered, but the riverine forests around that part of the Tocantins were full of them. I had seen several groups of them in the riverine forest adjacent to the hunting station. I had also seen a number of other rare species. Most spectacular among them were hyacinth macaws: massive purple birds, the largest of the parrot family and increasingly rare as a result of the pet trade.

I told this to the ornithologist as she returned from counting the local group of parakeets. She thought that the riverine forests in the area might be a staging post for the species during its annual migration, which would explain the high density of the birds.

She pushed plates, salt, and boxes of insects out of the way to write some notes, then walked over to the food and peered into another pot the cook had just brought in.

The cook was the final recipient of any specimens that came into camp, being last in line after the epidemiologists and the taxidermist. As in the most exclusive of London clubs, you never knew quite what would appear on the table, only that it would have a sauce over it. Even before I had left for the hunting station, armadillos, porcupines and other large rodents, howler, bearded saki, and white-fronted capuchin monkeys all put in an appearance at one time or another. The last was particularly unpleasant.

During my absence, the cook had surprised everyone, however, by displaying distinctly human emotions. He had befriended a large and extremely spiky iguana that would come into the back of the kitchen for scraps. Unfortunately, a collector who had arrived on the same boat as the ornithologist, had not been told in time and the iguana now reposed in a tub of formol. The cook was not happy and had been sharpening his cleaver in a threatening manner. That was the end of his flirtation with the rest of humanity.

I went to get some water from the tank, the filters being empty.

"Watch your fingers," said the herpetologist, collecting the tenstep and its bucket. "There's a snapping turtle in there."

I recalled that when I had left it had been piranhas.

From beside the water tank I watched curiously as a Gordian knot on two legs staggered past the kitchen en route to the laboratory tent. The herpetologist, tenstep, and bucket also came to watch.

"Is that you?" he asked eventually, perhaps recognizing his colleague's feet. "Need some help with that boa constrictor?"

For the next few weeks I wanted to stay around the main camp to work in forests on the west bank of the river, accessible by motor canoe. In this part of the Amazon basin, the terra firma forests were tall but uneven, with many gaps containing lower, almost impenetrable vegetation. Most of the large emergent trees, rising above the rest of the canopy, were Brazil-nut trees, whose cannonball-sized fruit littered the ground underneath them. ("Brazil-nuts," as sold in the supermarket, are actually seeds: each fruit contains many of these, surrounded by a hard outer casing and weighing about three-and-a-half pounds. Not something you want to fall on your head.)

The undergrowth in the forest was rich in spiny palms and various shrubs with associated ant colonies. I was constantly amazed at how many ants there were in Amazonia compared with elsewhere in the tropics. And some had extremely painful bites. Cutting trails through the forest I was constantly bitten by ants which swarmed over my arms as I pushed vegetation aside, or dropped down my neck from overhanging branches. There were plenty of wasps too.

Professionally, I was more interested in larger animals, particularly the primates. We had already found the rare bearded saki monkey on the east bank and wanted to compare population densities on either side of the river. This monkey is medium-sized and rather skinny underneath a coarse fur coat, and has a curiously shaped head with large bumps on the top of the skull. This is caused by an eccentric jaw musculature, necessary for the job of cracking the hard seeds on which the monkey specializes. Working with them wasn't very easy, however: they are very wide-ranging animals and were encountered only rarely during the surveys. We started to accumulate some information, however, and began to notice morphological differences between the populations on the different banks of the river. We failed to find the even rarer white-whiskered spider monkeys, which supposedly reach the western bank of the Tocantins. Of the numerous hunters we interviewed, only one remembered seeing them some way inland twenty years previously. Not a common animal.

Working out of the camp was more difficult than sitting out in a hunting station. Transport across the river was not always as punctual as might be desired, for one thing. However, the chance to socialize a little at breakfast and dinner was welcome. Conversation in the kitchen tent was invariably interesting and often enlightening. In the spirit of camaraderie through shared hardship (capuchin monkey, for instance), you sometimes learnt the secret dreams of your fellow scientists.

The leader of the team of epidemiologists came into the kitchen one evening and sat down with a sigh.

"I had a terrible day," he said. "Nothing was diseased at all. And I had a terrible night last night too. Partly due to that capybara and partly due to my recurring nightmare."

The reference to a capybara arose from a sighting of that largest of rodents on the edge of camp the previous midnight. The beast, which is the size of a large pig, had been busily getting its distinctly roman nose into the vegetables outside the kitchen. But the cook, ever protective of his onions, had spotted it. Visions of capybara steaks doubtlessly foremost in his mind, he had screamed a warning.

Never have I seen scientists out of their hammocks so fast. First into the swamp were the cytogeneticists: the epidemiologists and herpetologists tied for second. By the time I had extricated myself, being less familiar with hammocks, the swamp was half-full of scientists clad in various brightly colored items of underwear. All were splashing around in circles, shining lights into one another's eyes and shouting incomprehensibly. The capybara escaped. One of the herpetologists managed to grab a largish alligator, or vice versa. The entomologists slept through it.

"In this dream," the epidemiologist continued, "I've just died. I go through the Pearly Gates and there, in front of me, are rows and rows of animals. Big ones, little ones, adult ones, baby ones. Hundreds of them. All the animals we've killed for leishmaniasis screening over the years. They're all pointing at me and screaming 'It was him! It was him!' So down I go, down into the pit. The eternal torment. . . ."

He shrugged and peered into the stew, a haunted man. He ladled out a few spoonfuls of his latest victim.

As we progressed through the time allotted for our research efforts, we moved slowly south through the lake area. The forest varied little, but the floods that forced us to wade chest-deep through muddy water at the start of our morning surveys, holding binoculars and notebook over our heads, gradually receded. As the waters went down, vast expanses of blinding white sand were uncovered. To get from the access points of the river into our study areas we then had to spend hours baking in the reflected heat, following the meandering sandbars towards the terra firma forests. It would have been much quicker to wade through the many inlets that cut the sandbars into a mosaic of inter-connecting strips, but having noticed the three-foot-long, blue-spotted stingrays that buried themselves in the fine sand in the deeper water, I was disinclined to try. So we picked our way through the sandbar maze, eyes almost closed against the glare, disturbing the nesting terns, and frightening groups of black skimmers.

Over time, we began to get clear estimates of the numbers of animals that would be displaced by the formation of the lake. The numbers were considerable. We did not know, of course, how many would manage to swim from the flooding area to the margins, or how many would persist on the many small islands that would remain in the lake. But it was certain that many would drown, and also that overcrowding at the margin and on the islands would eventually lead to the starvation, or death through fighting and other stresses, of many more. Only predators could be expected to benefit, and even these, only in the short-term.

That one of the last remaining large patches of forest in the region should be lost through the flooding, and through associated developments, was tragic. But Brazil needed the hydroelectric power to support industry to generate foreign currency to service its foreign debt. Western banks financed the construction of the dam, perhaps to ensure the growth of the industrial base that would support the debt service.

Global politics was far above us, however, as we thrashed about in the undergrowth amidst the monkeys and herds of peccaries. Our role was to document the process. Perhaps our findings would be relevant to the many other hydroelectric projects that were being planned at that time. There had been many plans for harnessing the hydroelectric generative potential of the Amazon, from an international think-tank's scheme to dam the Amazon itself east of Manaus, to Brazil's own "Plano 2010," which called for eighty separate dams flooding a total area of 39,000 square miles. (Sounds like a lot, but that's actually only two percent of the extant forest area.) Under that scheme the Tocantins would be spanned by seven large and twenty smaller dams, turning it into a chain of lakes 1,200 miles long. Development on a large scale indeed.

Our final camp was located on a large island, some six miles long and three wide, in the middle of the Tocantins River itself. It forced the entire flow of the river into two sets of impressive rapids, in places only twenty to thirty yards wide and most exhilarating to navigate in a motor canoe. I spent the last few days sitting amidst the sandflies above the beach writing up notes. We had documented several variations among the monkeys, one of which would later be separated into a new subspecies, extended the known geographical range of several large mammals and located important populations of several rare birds. Various colleagues had discovered a new species of snake and a vast number of new insects, especially ants. Important reference collections had been made. Bales of skins and crates of bottled snakes and pinned insects were heading for the various museums around the world that would work on them. It would take many years to sort out precisely what had been discovered. By that time, the Tucuruí forests would be long gone, of course.

The expedition drew to a close with the pooling of data and last-minute collection of interesting animal groups.

"Pomarine ants," said the head of the entomology group informatively.

I had arrived back from the forest just in time for lunch and had thought that the plastic box on the rough plank table had something edible in it.

I slammed the lid back on, stopping most of its centimeter-long biting inhabitants from swarming out onto my rice and beans. Since the collecting was now largely complete, the kitchen fare had become less varied.

A bit further down the table an opaque plastic bag was revolving slowly around a tin containing spare knives and forks. Its curious perambulations suggested a bird-eating spider, but I didn't like to ask.

A herpetologist came into the kitchen tent for a glass of water and began sprinkling it over a small knot of gently writhing snakes lying gasping in a tin tray.

"Heatstroke," he explained as he walked off with them.

"Pomarine ants," the entomologist continued, "are most interesting creatures. Unfortunately, their taxonomy is very confused. The nineteenth-century taxonomist made a poor job of it, even by the standard of the times. He would forget where he was in the trays, come back after lunch and reclassify whole series of ants again in a completely different way. Hopeless. The group has just been revised extremely well, based on features of coloration of the thorax and abdomen. They fall into distinct groups. A divergent specimen has never been found, so the keying out of specimens is quite easy."

A small, undistinguished-looking pomarine ant which had crawled up onto the table from near the entrance to the tent chose that moment to make a dash for the sugar bowl.

"Take this specimen for instance," the entomologist said, picking it up and looking at it through a hand lens.

He put it down again and squashed it with his coffee mug.

Outside, various scientists were playing a game which involved guessing the weight of a boa constrictor that had drowned in one of the fishtraps and which went off the end of the herpetologists' scales.

"I want an accurate guess," one of them was saying belligerently. "This is scientific research, not a fairground."

The following day they collapsed the kitchen tent. That was the effective end of the expedition.

The Tucuruí dam was closed at the end of September. Our work in the area had lasted about seven months. We handed over to the animal rescue team. "Operation Curupira" swung into action, complete with helicopters, international film crews, and medical teams to deal with the people who grabbed the wrong sort of snake. Between September and the following April, some 20,000 monkeys were fished out of the rising waters, as were sundry other beasts, including over 25,000 bird-eating spiders. Some thirty million U.S. dollars were spent along the way.

Most of the animals drowned in the lake, nonetheless. Even the 470 people involved couldn't cover the whole area. Most animals that were captured were released on islands or at the lake margin, where most probably starved or were killed by the hunters who flocked to the scene. Generally unnoticed amidst the excitement, that particular piece of the Amazonian rain forest slowly disappeared.

No one fished any curupiras from the lake. Perhaps they can swim.

Siete de Enero

by Margaret Symington

Cocha Cashu is a wonderful place to do fieldwork. The field station's remote location in the Amazonian headwaters of southeastern Peru has thus far protected it from the ravages of "development" and the wildlife is tamer and more abundant than anywhere else in South America. The pristine rain forest surrounding the station veritably teems with species that are rare or endangered elsewhere in their range: jaguars, pumas, ocelots, tapirs, peccaries, giant anteaters, over 500 species of birds and, as one slightly jealous ornithologist had described it, so many monkeys they practically drip from the trees. For a field biologist, the place is no less than idyllic. But I needed to get out.

I was in my third year of a long-term study of Cocha Cashu's spider monkeys, and five months into a projected twelve-month stay. I was tired of getting up before dawn every day; I was tired of pulling on the same jeans and field shirt; I was tired of living with the odor of insect repellent in my nostrils; and I was particularly tired of eating beans and rice and tuna noodle night after night. I wanted to sleep late, put on a dress and some perfume, and go out to dinner in a nice restaurant, preferably in the company of someone I hadn't been sitting across from every evening for the last five months. So I decided to take a short break from my work and make the journey to Cuzco for Christmas.

Since the trip each way would be at least five days, this was not a decision to be made lightly. In fact, it would have been difficult to justify the effort and expense of such a trip on the basis of rest and relaxation alone, but food supplies at the station were also running low and a supply trip was in order. When the ratio of weevils to rice reached a certain level, it was no longer a matter of taste, but also of hygiene.

When New Year's Eve had come and gone, I turned my back on the bright lights and fine restaurants of Cuzco and prepared to return to the field. I had never made the trip from Cuzco to Cocha Cashu in the wet season before, and I was not looking forward to it. The trip was rugged enough in the dry season. The first stage consisted of sixteen to thirty-six hours in a truck on a winding Andean road so narrow that eastward traffic was restricted to Mondays, Wednesdays, and Fridays and westward traffic to Tuesdays, Thursdays, and Saturdays. Only those with a serious death wish would venture onto the road on Sundays, when the traffic went both ways. If you survived the truck trip (the roadside was dotted with small white crosses inscribed with the names of those who had not), you then boarded a twenty-five-foot, motorized canoe for a leisurely ride lasting three to five days depending on the horsepower of the motor and the height of the water. In the wet season, heavy rains could further complicate things by washing out roads and flooding rivers.

We bought our supplies from a grocer in Cuzco; enough to feed six people for six months. Fifty pounds of rice, twelve pounds of beans, twenty-five pounds of noodles, six cases of tomato paste, three cases of tuna, forty-five pounds of sugar, twelve pounds of flour, eleven pounds of powdered milk, twenty-five gallons of vegetable oil, two cases of canned green beans, one case each of canned peaches and pineapple, six large tins of soda crackers, eight family-size jars of strawberry jam, 100 rolls of toilet paper, ten pounds of oatmeal, four pounds of Nescafe, seven cases of beer, twelve bottles of rum, laundry detergent, candles, and matches. We also went to the open air market down by the train station and bought as many sacks of onions and potatoes as we thought we could eat in the short time before they rotted or were consumed by the family of mouse opossums that shared our cookhouse. These ingredients and the limited number of permutations possible therefrom would comprise our entire diet for the next six months.

The group going into Cocha Cashu this time was a small one: me, my Peruvian field assistant, Fernando, and an enthusiastic American tourist, Jo, who had visited the site the year before on a luxury nature tour and had liked it so much that she wanted to come back and spend the wet season helping me out with my research on spider monkeys. Although our group was small, our supplies were bulky and I arranged to charter a truck from a driver we frequently worked with. The truck was a classic '56 Chevy pickup with wooden side rails built onto the bed to permit larger loads. By the time we had loaded our supplies, ourselves, and the three fifty-gallon drums of gasoline we needed for the boats back at Cocha Cashu, the pickup was nearly full, if not downright cozy.

Although we had planned to leave as early in the day as possible, a heavy morning rain delayed us. Our driver took advantage of the delay by getting outrageously drunk. We left after lunch despite his condition, since we knew if we waited any longer we would never make it to the checkpoint, where they regulated the direction of the traffic, by the required cut-off time. The only vehicles allowed to pass were those that had a reasonable chance of making it across the Cordillera by midnight, and if we didn't get there soon, we would be stuck.

The delay was lucky in one way since we ran into Juan, who was on his way into the park to work on the construction of another field station only a few miles away from Cocha Cashu. Juan was a large, extremely strong fellow, well-suited for the construction business. Since his only luggage consisted of one small bag for his personal possessions and a brand-new chainsaw that he had been charged with escorting to the construction site, we welcomed him into our group.

I remember little of the trip until we reached Pilcopata, a largish market town and truck stop on the eastern side of the Cordillera, well on the way into the Amazon basin. We rolled into town just before dawn and pulled over to the side of the road to join the dozens of other trucks full of lumber, bananas, rice, and sleepy *campesinos.* The jungle is a frontier in South America, and Pilcopata was a frontier town. Loud drunks, singing songs from their highland homes, often wandered the street until sunup, and violent machete fights were not uncommon.

I was somewhat relieved that we had arrived in Pilcopata too late to get rooms at the local pension. The hygiene there was questionable and I was satisfied (although not overly comfortable) with the bed I had fashioned from a twenty-five pound sack of rice. Although one could doze while the truck was still moving, the constant bumps and jolts made it impossible to really sleep. We all (including the driver, who had been waiting a long time to sleep it off) fell into welcome oblivion until the first roosters began to crow about an hour later.

After eating a hearty breakfast of eggs, rice, and yucca, liberally doused with *aji,* the local hot pepper concoction, we piled back into the truck for the worst part of the ride: crossing the broken-down bridge over the Pilcopata River. Rushing white water and jagged boulders were easily visible between the wide gaps in the two-by-fours which paved the bridge. It was morning rush hour in Pilcopata, and the trucks were lined up on both sides of the river, waiting for their turn to cross.

As each truck crossed, the passengers would climb down from their perches and precede the truck and its cargo across the roaring river, their most precious possessions clutched in their arms. When it was finally our turn, we stood and watched the driver cross himself and kiss the rosary that hung from his rearview mirror before we turned our backs and crossed the bridge. Fernando carried his brand-new sneakers that he had gotten for Christmas, Jo carried a backpack full of special camp food (chicken stew packed in vacuum sealed aluminum pouches) that she had brought to supplement our spartan vegetarian supplies, Juan carried his chainsaw, and I carried a small waterproof bag filled with the original copies of the data I had so laboriously gathered over the last five months at Cocha Cashu.

Although the rest of the truck trip was uneventful, it was more than ten hours before we reached the end of the road, where it ran literally headlong into the Alto Madre de Dios River. Shintuya, the small mission town located at the road terminus, was a picturesque town with a beautiful view of the Sierra de la Pantiacolla, one of the easternmost outcroppings of the Andes in Peru. In the mornings, a line of clouds would lie like a fluffy scarf on the shoulders of the sierra. From here, all the way to the Atlantic coast of Brazil over 3,000 miles away, the land would drop only 1,500 feet in elevation. In between, lay the vast rain forests of the Amazon. With any luck and the appropriate boat, we would be safely settled within that vast forest at Cocha Cashu in a couple of days' time.

We unloaded all of the supplies from the truck and placed them under a makeshift lean-to that was used to keep timber dry as it waited for a truck to take it from the jungle to the market. I had permission to use my Ph.D. adviser's fifty-five-horsepower outboard motor which he kept in storage in Shintuya, and we had brought our own supply of gasoline, so all we needed now was a boat. Edmundo, the owner of the local restaurant/bar, assured me that the only boats likely to come to Shintuya this time of year were the Moscoso's. He thought that one of them (the family consisted of a father and six sons) would show up very soon, if only to buy beer. Then again, and Edmundo shrugged, no one might come for another week.

Although I was sure there were worse things possible than being stranded in Shintuya for a week, it was hard to think of them at the moment. So we slathered on the insect repellent (Shintuya was notorious for large quantities of biting black flies) and sat down on our backpacks to watch the river flow.

It turned out that we didn't have too long to wait. The next morning around ten, the high-pitched hum of an outboard motor made itself barely audible above the hum of black flies. About ten minutes later, Carlos Moscoso and his younger brother, Gustavo, pulled up to the bank below us and bounded ashore. I quickly approached their boat to begin negotiations.

Although Carlos was traveling on to Pilcopata to take care of some business for his father, Gustavo would be glad to take us all the way to Cocha Cashu. Gustavo was not as experienced a *motorista* as Carlos, and had quite a reputation as a ladies' man besides, but neither point really worried me too much. The latter might even add a little excitement to the trip. Of course, there were a few minor complications. The boat they were in was loaded with wood, destined to become railroad ties in some more developed part of Peru. This would have to be unloaded from the boat and loaded onto the truck we had come in. And once we reached Gustavo's house, we would have to switch boats since this one leaked and, besides, was much too small to carry all of our gear safely. But in principle, the deal was done.

By the time all of the unloading and loading had been accomplished, and Carlos and our trusty driver had receded back down the road to Pilcopata, it was two o'clock. We set off in high spirits. It felt great to be on the river; the water was high, the current was fast, and the boat was definitely overloaded. But Gustavo knew the Alto Madre well and it took us less than an hour to reach his house.

Transferring six months' worth of provisions between boats was no mean feat and I gave silent thanks for Juan's presence, as he practically single-handedly lifted the three fifty-gallon drums of gasoline into our new boat. Fortunately, this boat was substantially larger and less leaky than the first one, but it was still a tight fit to get all of our gear situated securely so that it wouldn't slide off into the river if the boat happened to tilt. Around four, we set off once again.

Twilight on the river is a beautiful time. Macaws traded overhead, giving raucous, prehistoric cries as they headed toward wherever it was they passed the night. Palm trees, with the sun low in the sky, took on resort brochure silhouettes, and flowering vines of purple and gold stood out from a canopy that was a thousand shades of green. The air turned cool and the black flies disappeared, but it was still too early for the mosquitoes. With every river bend we rounded, there was the anticipation of seeing a capybara drinking at the water's edge, or perhaps a tapir, or maybe even a jaguar. The difficulties of the trip seemed insignificant at moments such as these.

The beaches on the Alto Madre are rocky, but as if by divine providence, the next one we passed after the sun sank below the horizon had a large patch of soft sand that was perfect for setting up tents. Gustavo made a wide U-turn and cut the engine to bring the boat up alongside the beach.

Fernando, Jo, and I hurried to put up our tents before dark, while Juan gathered wood for a campfire and Gustavo tended to the boat. A sense of peace pervaded, and Jo and I didn't even mind that the men seemed to assume that we would cook dinner. We made a big pot of noodles and topped it with a couple of pouches of Jo's special chicken stew. About the time we finished eating, the mosquitoes came out in force and we all retreated to our tents for the night.

Just after sunrise, I heard the familiar zip of a tent nearby and quiet rustling as someone got up. A minute later, the rustling intensified, a second tent unzipped, and I heard voices conversing in Spanish, worriedly. Through the fog of sleep, the phrase "*completamente barrada*" made me bolt upright. I quickly pulled on my pants, unzipped my own tent, and stuck my head out. Gustavo and Juan looked at me sheepishly, and I turned toward the boat to confirm my worst fears. Sure enough, the boat stood high and dry, with nothing beneath its hull but rocks.

Gustavo tried to explain. He had no idea that the river would go down so much overnight. A few feet maybe, but six feet! It was outrageous! He hadn't thought it necessary to get up in the middle of the night and check on the boat. I smiled weakly, and kicked myself for not having thought of it either. It didn't matter now: we had a tough morning of work ahead of us.

Fernando and Jo were quickly roused and for the third time in less than 24 hours we went to work unloading the boat. When we got the boat half unloaded, we all pushed and rocked and huffed and puffed and managed to get it back afloat. After a short break for breakfast, we reloaded it with the remaining supplies and were back on our way.

Midday on the river is a very different experience than late afternoon. The sun beats down with a ferocity that is almost unimaginable. All living things withdraw into the cool shade of the forest and the hum of the outboard motor echoes down the empty river. I donned my safari hat and sunscreen and entered a sun-induced stupor.

Some time in the early afternoon, we arrived at Boca Manu, where the muddy brown Manu River joins the clear blue Alto Madre. For a time the waters run side by side, separate and clearly discernible. Then they blend to form a third distinct shade, clearer than the Manu, but darker than the Alto Madre. From here on, our trip would be upriver, not down, and with many more bends and submerged logs. It was a completely different river and one with which Gustavo was not nearly so familiar; and from the very first moment, it looked like trouble.

Gustavo's brow was furrowed and his full attention was focused on the roiling brown waters ahead. He looked at me and shrugged. He said that the Manu looked to be in full flood. The way he said it made me nervous. He opened the engine full throttle and pulled us in tight to the riverbank. I stared at a small sapling about five feet away on the bank and noticed that it wasn't moving. *We* weren't moving. The current was so strong that even with a fifty-five-horsepower engine, we were standing still. I shot Gustavo a questioning look. He shrugged again and said I should be happy we weren't going backwards. After a few more minutes of useless effort, Gustavo decided we would have to give way. Since it would be impossible to turn the boat around in water like this, he simply cut back on the power and we drifted backwards into somewhat calmer water.

Just then, I noticed a large tree trunk coming down the river. A yell came up from somewhere ahead of us and a small canoe, powered by a sixteen-horsepower Briggs and Stratton motor (known as a *peque-peque* in the local parlance for its staccato intonations), bobbed out into the turbulent river. The motorista was accompanied by a single passenger, who stood, George Washington-like, in the bow of the boat. They were heading straight for the trunk, and when they were within an arm's length of crashing into it, the man in the bow reached out and slipped a noose around it. Once the end was secure, he roped in the other end and pulled the trunk alongside the canoe so that it resembled an outrigger. During this whole process, the motorista had been skillfully maneuvering to keep the canoe in the proper position, facing upriver, and holding steady against the current. Now he turned the canoe back down the river, with trophy in tow, and pulled alongside us.

Gustavo congratulated them on their catch, and said that it would make a fine canoe someday. The two men agreed, and added that it certainly must be some flood upriver with all the mahogany trunks that had been coming down. Gustavo told them that we were trying to get upriver to Cocha Cashu, but thought that we would have to wait since the current was so strong here that we couldn't get through. The men replied that if we really wanted to get through today, all we had to do was go through the old river channel, which was usually dry but had plenty of water in it now.

River courses change frequently in the Manu basin, but this particular change had been hard on the small community of Boca Manu. Many of the ten to twelve houses that made up the village now faced out on a mosquito-infested backwater during much of the year, rather than on the main river channel that was the source of their livelihood, as well as their only connection to the outside world. Some had plans to move their houses; others planned to wait and see if the river would eventually come back to its old ways. But the old channel served our purposes perfectly.

When we came back out onto the main channel, the current was fast, but at least we were able to move against it. Gustavo had to hold tight against the riverbank, where the water was least turbulent, crossing from side to side to stay on the outside bank. To travel along the inside bank would be dangerous, since that was the cutting edge. With currents this fast, you never knew when a large chunk of riverbank and assorted debris might come off and get in your way. Switching from side to side was both time-consuming and dangerous since the river was swollen to about three times its normal width and, while crossing, one was exposed to drifting trunks and large branches, which always came down the middle of the river.

We settled into a groove though and, before long, two hours had gone by. Our progress was extremely slow, and Fernando and Jo, who were seated amidships, both appeared to be asleep. Juan, who sat in the bow, was our lookout for trunks, and was diligently watching the center of the river to tell us when it was safe to cross. I sat in the stern across from Gustavo, chatting amiably and admiring his unusual and attractive green eyes.

"Watch out!" Jo's voice rang out.

None of the Peruvians in the boat spoke English, but the panic in her voice was universally understandable. All heads were up, and we watched as a large tree on the bank ahead of us, silently and as if in slow motion, fell directly on the bow of our boat.

I gasped, and then swore, as the boat continued inexorably forward. It dragged us through the thorn-covered branches of the tree, until we finally ground to a halt with the tree laying directly amidships. There was a moment of shocked silence and then Juan's head popped up from the river right next to me. He was all right: he had jumped off the bow just a second before the tree had hit. I checked everyone else and heaved a sigh of relief; apart from some dramatic bleeding cuts from the thorns everyone looked okay.

Before we even had time to think of what to do next, Jo piped up again. "Uh, I think there's water coming in," she said in a much quieter, almost resigned tone. Sure enough, there was water coming in, lots of water. In fact, we were sinking.

The next few minutes were confusing and scary. The tree had cracked our bow, and we were taking on water fast enough to sink us within a few minutes. Gustavo exhorted us to save beds and food, and Juan, bless his soul, was strong enough to throw some of our personal gear, which was arranged on top of the supplies, safely to shore. I watched with relief as the backpack that contained my data landed with a thud on top of the muddy bank. Juan pulled desperately at the box containing his precious chainsaw, but was thwarted by the twenty-five-pound sack of rice that lay on top of it. As the boat sank deeper, provisions—I remember vividly the orange plastic bottles of cooking oil—began to float away down the river. When Jo's pack filled with chicken stew began to do the same, I jumped overboard to save it. The strength of the current was frightening. I briefly let go of a nearby tree branch in order to grab the pack of food, and felt myself being swept down the river. Although we were close to the bank, the water was deep, over our heads, and any sort of salvage was impossible as soon as the gunwales of the boat sank beneath the muddy brown water.

When the boat sank, it flipped over, sending the rest of our supplies either directly to the muddy bottom or bobbing down the river like a long multi-colored parade. We sat dazed, either squatting on the muddy bank, or still in the river clinging to the fateful tree. Suddenly, I remembered the motor. We had to save the motor. If we didn't, I would be washing dishes to pay for it from now until the next century. Luckily, motors are always tied to boats with a length of rope. Gustavo decided that the safest thing would be to tie the motor to the fallen tree, which was more securely anchored at that point than the boat. It turned out to be a wise decision; less than an hour later, our shattered boat was swept away by the still rising flood waters.

I can't remember exactly when we started to laugh. Maybe it was when we began to recount all the little things that we had lost: Fernando's sneakers, my safari hat, Juan's favorite blanket, Gustavo's cigarettes. Or maybe it was when we discovered that only one tent had been saved, and we set it up, and all piled inside, five soaking bodies in a two-man tent. Or maybe it was when we started eating Jo's chicken stew right out of its aluminum pouch. But it probably was when we discovered the bottle of Scotch that Jo had hidden at the bottom of her backpack, and proceeded to pass it around the tent until it was empty.

Although we joked a lot about being shipwrecked and spending the rest of our lives living like savages on the banks of the Manu River, we knew it was only a matter of time until a boat went by that we could flag down. With a tent to protect us from the mosquitoes, a decent amount of food, and enough fresh water to float an ark, the only real danger we faced was discomfort. Since we were only three hours or so upriver from Boca Manu, we all agreed it was likely that someone would have seen our supplies floating by, correctly assumed that we had had an unfortunate accident, and would come to investigate. Boat accidents, although infrequent, were certainly not unheard of.

Sure enough, the next morning, about two hours after sunrise, a caravan of three canoes rounded the bend coming from the direction of Boca Manu. The boats were packed with men, women, and children; it looked like half the village had come out for the excursion. We waved from the shore and, cheerfully, they waved back. The women had brought fruit, freshly picked mangos; the men had brought ropes and tools. They told us how they had seen an orange plastic bottle of cooking oil, then another, and another. Then, when they had seen the drums of gasoline go by, they knew. We told them how it had happened; the men shook their heads, the women and children laughed. The infinitesimal probability of being hit by a tree falling from a riverbank was no longer relevant. It was fate; it had happened.

A team was organized to retrieve the motor from the bottom of the river. After substantial effort, the motor, dripping fine silt from every pore, was dragged up and loaded onto one of the boats. It would have to go back to Boca Manu for a complete overhaul, to be taken apart completely, cleaned, and then reassembled. This would have to be done soon, if we wanted to prevent it from rusting into a muddy monolith.

In a touching gesture, the people of Boca Manu announced that the small stream that entered the Manu across the river from the site of the accident would, henceforth, be called *Siete de Enero* for the date on which the accident had occurred. I was to take many more river trips up to Cocha Cashu, and I would always ask the motorista to point out Siete de Enero. They always knew; this was probably because most of the residents of Boca Manu returned many times after the flood waters had receded to salvage as much as they could from the muddy bottom of the river. People passing through Boca Manu shortly after the accident related strange tales of bushes festooned with drying toilet paper, and unusually abundant supplies of canned peaches and rum.

Since Fernando, Jo, Juan, and I would continue upriver to Cocha Cashu, and Gustavo would go back to his house to attend to the motor, we said our good-byes. Gustavo and I shared a special camaraderie from then on, and although he was never my motorista again, it wasn't because I held the accident against him. Others said he had been careless, and should have crossed to the outside bank sooner, but all of us who had been there on the Siete de Enero knew that the accident hadn't really been anybody's fault. It had been fate.

I was to spend the next seven months at Cocha Cashu without a break. Up before dawn, and out watching monkeys until dusk, I now generally found the routine to be comforting rather than oppressive. But whenever it got to be too much, like after the fifth night in a row of noodles and tuna, I would simply glance at one of my water-stained data books, and the urge to get out would disappear.

Jungle Love

by John Symington

I WENT TO work in the jungle for the woman I eventually married. I figured that the jungle would be a wonderful place for passion to flower and that the headwaters of the Amazon in Peru would be the perfect setting to clarify romantic inclinations. I envisioned three months of nature loving, cavorting with an attractive girl, and lots of delicious, fresh, tropical fruit. What I discovered instead was that it was incredibly hard work to keep up with a primatologist in her element. I also found that field biologists eat mostly canned fruit or occasionally something someone's study animal had been eating which was invariably quite small and often bitter or tasteless. As a medical student, I arrived at the field station of Cocha Cashu with few bush skills and even less expertise in primate behavior. But in the span of three months I was capable of recognizing virtually identical looking monkeys as individuals; I learned how to cut a decent trail with a machete without dismembering myself, and I was able to find the way back to my tent in the dark. Of course, I also fell in love.

Ateles paniscus chamek was the species of monkey we studied. Although commonly referred to as the spider monkey, it was also known as *maki sapa* in one of the local dialects. This means "big hand," a name bestowed for the extraordinary length of its four fingers (it has no thumb). Groups of these monkeys abounded at Cocha Cashu where the study area was nestled in a bend of the Manu River within the Manu National Park of southeastern Peru. Twelve miles of trails crisscrossed the 140 acres of forest and swamp that fell between the two cochas (or oxbow lakes) that had been made by natural changes in the course of the Manu River. The wildlife was astounding and included eleven species of primates, over 500 species of birds, and more than a thousand species of trees. Unfortunately mangos, papayas, and other tropical comestibles were not in evidence as there were rules against having introduced species at Cocha Cashu.

Perhaps the most memorable denizen of the jungle was my little flying friend, the mosquito. There were so many of them that I found myself actually talking loudly over their high-pitched crescendos and decrescendos. They would cover my body completely, inspecting, probing, and prodding until they were satisfied that I had not failed to douse any exposed skin liberally with insect repellent. The scalp has exceptionally good blood flow, more so than most skin on the human body. The little devils would often line up like pigs at a trough where I part my hair, knowing that was the place where I had not applied repellent well. When areas of exposed flesh had been exhaustively researched, these flying phlebotomists would alight on bony prominences in hopes that a square meal could be obtained through clothing.

Chances were improved if their victim failed to wear an undershirt, which increased the distance of cloth beyond that of the mosquito proboscis. And as a diet of mainly beans and rice caused this field assistant to go from 155 pounds to 142 pounds in three months, there were more and more bony landing strips on my body as the field season wore on.

But these things paled in the bright cheer of growing affection. The whole adventure in fatigue and spare living only served as a catalyst in developing the friendship between me and my boss, the primatologist. I should say that two people contemplating a serious relationship should begin by taking an extended camping trip together. In this way they can discover just how compatible they are when there is only a single slice of bacon or one chocolate bar left, or when it continues to rain for the fourth day in a row. A good, long, eventful camping trip can telescope years of experience and expose character better than any method I know. So, I began to like the perky redhead who gave me my marching orders each morning, and made sure my pants were tucked into my boots to avoid ticks and other vermin.

Feeding time was generally pleasant for humans as well as for mosquitoes, if one didn't concentrate heavily on the fare. Fresh vegetables lasted about four weeks. We ran out of carrots first, then turnips, then potatoes, and finally the last onion was eaten or rotted. The trick was to eat the perishable food slowly enough to make it last, but fast enough to avoid excessive spoilage. We often erred on the side of stretching out vegetables and ended up eating what you might call very ripe food. Eventually beans and rice became the major source of nourishment. My time in the field was relatively short, so it seemed masochistically poor form to openly fantasize about favorite, inaccessible foods. Field biologists continually talk about foods that they would not eat for months. I diagnosed this behavior to be a food separation neurosis, and I pointedly restricted my conversation to the four food groups of the field (beans, rice, powdered milk, and coffee) in order to avoid exacerbating anyone's gustatory obsessions.

In the jungle, most days consisted of picking one animal (a focal animal) to follow throughout the day, and recording what it did every three minutes. Other days were spent measuring the dimensions and noting locations of trees used by our focal animals. In truth this job required little need for a medical background until one day when a spider monkey was found dead. Here was my chance to be useful. Spider monkeys were large and generally considered prey for only the largest of monkey-eating eagles. Other causes of death were not well known so a necropsy would be invaluable. Of course, this was before I knew about green monkeys and their connection with the human immunodeficiency virus (HIV) or about human deaths caused by exposure to monkeys infected with the Herpesvirus simiae. So I had no qualms about opening up the animal.

We found the young male lying on the jungle floor with no obvious sign of disease or trauma. In fact, he was in good condition with nice fur and good teeth. I brought him back to the clearing and inspected him more closely. Some flies had already laid their eggs and a few maggots were at the corners of his eyes, so he had been dead at least a day. He weighed fifteen pounds, which is respectable, and I couldn't find any fractures. So why did he die? I opened his chest cavity, which revealed a normal heart, and lungs. I then exposed the abdominal cavity, inspecting for hemorrhage or signs of parasites obstructing the intestines. Beginning with the duodenum and working down, I soon came upon the largest appendix I had ever seen. It was bigger than my fist. It was huge, especially for the size of the monkey's gut cavity. Now, I had just finished my anatomy course, and even though I had no clinical expertise in recognizing appendicitis, the shear bulk of this structure seemed to say it all for me. I opened the appendix, anticipating pus, when a matted ball of half-digested leaves and seeds fell out. Well, I reasoned, maybe he got septic from appendicitis and developed a paralysis of his intestines. Parenthetically, I should note that appendicitis was heavily on the minds of most researchers at Cocha Cashu, as it would take at least three days, maybe longer, to reach a surgeon if they developed appendicitis. For this reason, some researchers had actually considered elective appendectomies at home prior to doing fieldwork at Manu. During the evening meal, I made a solemn report of my findings to the group and, after a polite pause, I was quietly told that spider monkeys ferment leaves in their appendix, which was normally quite large. At the first light of morning I got up to reinspect the dead animal. I prepared myself for the challenge of a decomposing body but to my great fortune the monkey had been taken away in the night. Whew! Now I would not have to suffer through another stinky necropsy and risk losing face a second time by making another misdiagnosis. I wondered how he died, but I thought even more about what animal was big enough to drag him off in the night.

One day we chose "Stumptail" as our focal animal. He was a juvenile male who had lost his tail and was left with only a stump. This put him at a severe disadvantage when traveling through the canopy, but on several occasions we saw both adult females and males spanning the distance between two trees with their tails and arms so that Stumptail could walk over their backs from one tree to the next. No other monkey his age was given this kind of treatment and it probably allowed him to live as long as he had. Stumptail enjoyed play and we often saw him in the company of spider monkey babies, which are by far the funniest animals to watch in the jungle. These little, stiff, yet wobbly, creatures would grope about their mothers' bodies and attempt various escape maneuvers, but their mothers always managed to grab a tail or leg in the nick of time. When the babies were a little older, they would hang upside down by their tails, which was spooky, because they seemed so uncoordinated; but there they would be, dangling with a three-hundred-foot free fall to the jungle floor below.

It was a pleasure following Stumptail, because he was entertaining and could not race through the trees and lose us. One day he was overhead, ambling about a newly fruited fig tree in the company of two adult females when we heard a crash to the south. Two large spider monkeys were heading our way. From the amount of noise they made, they were probably going to be males and indeed they were, coming in a fury, arms reaching out to grab limbs after impossibly long leaps between trees. Jumps by the heavier males produced the loudest crashes when they took off from a tall tree and then sailed down, legs spread wide, onto a leafy, neighboring limb. These fellows were getting a lot of mileage out of this technique, moving quickly through the forest and with loud crashes they let the world know about their big, daring bodies.

The two females gave several high-pitched chortles and Stumptail began to exit the tree. Both males came into the tree shortly after Stumptail left, in hot pursuit of the little handicapped monkey. I was worried. We had taken a proprietary interest in Stumptail. He was practically our son. I knew that they would easily catch him, and then what? Would they hurt him? Would he fall trying to get away? Fortunately, the two males stopped short in their chase and ended up in a small tree right over my head. And then it happened: Like great athletes in sport, the two satisfied hellions, having proven themselves, turned towards each other for recognition. They were both standing on the same branch thirty feet above. The tree was small, about six inches in diameter and the branch swayed gently under their weight. The one on the left had his arms stretched out in front of him with his palms upturned. The other moved towards the first and when he came within striking distance, his hands came up and then down, and slapped the upturned palms of his friend. Then they immediately put their arms around each other in a hug. It was all too much for me. I had just seen two spider monkeys slap ten and I actually fell to the ground with laughter.

Annoyed by my reaction, the males grabbed nearby limbs and shook them to frighten me. My laughter slowed as I gazed up at these two bullies with their fur all puffed up in pugilistic piloerection. They looked upon me disdainfully: a silly ground primate. Who did I think I was, disturbing their revel? I got up and looked to the west, searching for Stumptail. Poor guy, bullied by these thoughtless brutes. I looked up at the dynamic duo still staring down at me, and was seized with the desire to set them back a notch. I walked over to the nearby tree and shook the whole thing. Instantly one of them screamed and threw himself into the arms of the larger one, who dutifully caught his friend while staring down at me. Moments later they scrambled up the tree and over into the fruiting fig where the two females chortled softly on their arrival.

Three months flew by and suddenly I had only a few minutes more to enjoy the company of my friends in the jungle. How I would miss those relaxing swims at dusk in the piranha-filled cocha. I would no longer experience the sense of morbid wonder generated by a dive into the water. Would I hit a turtle, bleed profusely, and not make it back to shore alive? Nor would I entertain the equally paranoid thought after being bitten by a sandfly that the bite would give me some parasitic infection like leishmaniasis? And then there were the ever present ants, how would I live without so many ants? But more importantly, how could I live without this woman whom I had grown to love and admire. I boarded the departing boat without her and stared out over the water, looking back at her as we pulled away from the shore. A lump gathered in the bottom of my throat as I perched on the side of the canoe. I waved somewhat feebly, feeling a sting inside my nose proceed the welling up of tears. Farewells were shouted out over the sound of the *peque-peque* boat motor, "*Vaya con Dios*," "See you soon," "Have a safe trip." In a few moments we were well out on the river and the people on shore were quite small. I looked forward into the wind and my thoughts turned to when I would see my friend again, and how I would ask her to marry me.

The Birth and Death of a Very Fine Pit

by Ronald E. Cole

ON A SCALE between one and ten, this field camp would not score well—two huts, hastily built in just a few hours from saplings, tied together with vines and covered with ferns. This is to be our home away from home for some period of time. How long we stay here remains to be seen. You see, we've run into a bit of a problem with the villagers living just down the trail, and for our own safety we have stopped our overland trek, taking up temporary residence at this small spot in this very large forest.

This is Papua, New Guinea, where the best of plans must constantly be changed, and where the expected should never be. Andy Engilis, Frank Radovsky, and I have been contracted by the Bishop Museum and the University of California to conduct a survey of the birds and mammals in the high mountains at the eastern end of this island, and the problems we've just encountered are not part of our plans. The logistics of this expedition are somewhat complex. With no roads in the region, travel is by foot, the trails are not mapped, we have a fair amount of collecting gear, and the people we've encountered, living in small, scattered villages along the way, are somewhat hostile. Violence is a way of life in New Guinea, and a constant threat to outsiders like us. Aggressive behavior coupled with little written information about the region and its people leads to many unanswered questions. Unanswered questions can get us into trouble. This assignment, while not impossible, will be challenging.

We encounter problems fairly early on. After catching a ride in on a bush plane and making arrangements with the pilot to pick us up at the same drop-off site some weeks later, weather permitting, we had hoped to meet with the elders of the tribal groups living between here and our chosen field site to ask for permission to work in their territory. Unfortunately, the elders don't want to meet with us. It seems there is some unrest, rival village feuding, and perhaps some long-held resentment toward outsiders, and we've stepped into the middle of it.

In New Guinea payback is a way of life. An act of treachery is inevitably met with retribution. To outsiders, like us, wanting only to walk from here to there, and to collect some birds and mammals for science, this is all too strange. While we respect customs, we really don't want to be involved in things not of our making. But we are involved, at least inadvertently. We learn that some years ago a team of "Europeans" came into the area collecting plants. They were probably collecting orchids, which grow on every tall tree in the cloud forest, and unless they were collecting for a museum, they probably made some money from their efforts. Apparently they didn't share profits with the local villagers, and over time this disrespect has festered in the villagers' minds. Now we come along, wanting to collect some possums and scrub wrens, and obviously we must be treacherous, also. While we may not deserve payback, we can't expect any special treatment either.

So we find ourselves caught in a slight squeeze. We can't walk north, to the mountains that we had originally called our objective, because to do so would rob the local villagers of their birthright spirits, or so we are told. We don't want to walk south, back in the direction that we have traveled, because we were dropped off in a valley where the landscape is fairly well disturbed by the village and the gardens that support the villagers. This habitat disruption insures that only the most commensal and universally known of species would be found in the vicinity of the village. To make matters worse, the villagers living in that valley are suspicious of us and our motives. Besides, the plane that brought us in is long gone, and won't be back for some time. We're stuck right here, on top of this ridge, just a two-day walk from where we want to be, but a quick death, we are told, if we go on. The only thing to do is tread cautiously, which we do.

To a field biologist, New Guinea is one of the great challenges. The topography is harsh. The lowlands are not hospitable. They are hot, wet, swampy, snake and mosquito infested bogs that stereotype everything you might have imagined about jungles. They are too wet to support roads, hence the forests remain largely intact. Because the swamps make foot travel nearly impossible, the people seldom see outsiders. Once you rise above the lowland jungles, you enter into the steep terrain of the Owen Stanley Mountains, also inhabited primarily by people who have had little outside contact. Attempts have been made to build roads into the lowland jungles and the mountains, generally without success. This has slowed down deforestation, but the timber resources are so large, and the potential profits from hardwood harvests so great, that it is only a matter of time before technology prevails. In the meantime, life goes on essentially as it has for centuries. There are over 700 identifiable languages spoken on this island. One of the reasons for so many tongues is a long-standing lack of communication between very warlike groups living in isolated valleys.

In gratitude to the leader of the family on whose land we are temporarily encamped, we hire his cousins to work for us. A motley crew on first appearance, these chaps turn out to be very special people: rural naturalists, amazing athletes, skilled bushcraftsmen, multilingual, and friendly. We know that they have tribal names, but they never tell us what these might be. Rather, we know them as Arthur, Stephen, and Faithful. Faithful is our favorite. He is just five feet tall, with rocksolid muscles, and while he looks like he just ate someone for dinner, he is remarkably pleasant. He rounds up about forty of his friends who carry our supplies and equipment into the bush, and after paying them a fair wage, and serving everyone a cup of hot tea, we thank and send them on their way.

If this is going to be our home for a number of weeks, we had better make it livable. First on our agenda is a place to sleep, a place to work, and a place for our helpers to sleep. Tents don't work in the tropics. They are too small, too heavy, too confining, and too expensive. Everything bad about a tent can be corrected with a hut of saplings tied together with vines. Materials are local; materials are free; the huts can be of any design and any size that will fit into the landscape; huts can be built in very short order. After we build ours, we construct a sleeping and cooking hut for our helpers. Next we build a work table where we'll spend long hours, preparing specimens, writing notes in our journals, and doing the other strange things that we, who know the secret to material life, do.

Once settled, we begin to collect and prepare specimens. We brought hundreds of traps and dozens of nets with us, which we set out daily. We also send out word that we are interested in buying specimens, and will pay big *kina* for certain animals. Luckily for us, every bird and mammal in the area has its own "place-talk" name. And even better, Faithful tells us that he knows *all* of the names. He has a different name for every rodent and every possum that we see, and the same holds true for birds. This seems almost too good to be true, and we decide to play a game with him to test this ability. One day he brings us a male Sclater's whistler and the game begins. This little bird appears quite often in our nets and is distinguished by its olive-green back, white throat, black chest stripe, and bright yellow breast. He tells us that this bird's place-talk name is *sissytautau.* The next day when he runs our nets, he brings us a female whistler. Same species, but quite different looking because of the sex difference. To challenge him we query: "Faithful, you told us that this bird from yesterday is a sissytautau, and now you tell us that this other bird is a sissytautau, but they don't even look alike. What gives?"

"One is a *meri,* and one is a man," he says. And, of course he's right. One is a female and one is a male, and Faithful probably wonders why we don't know that. We're impressed, but the best is yet to come. The next day Faithful brings us a juvenile Sclater's whistler, and the game heats up. Completely unlike either the adult male or the adult female, this little fellow is rusty-brown with just a hint of green color on its back. From all appearances, a different species entirely.

"Faithful, this is a new bird for us. What do you call this one?"

"A sissytautau," he says. Ah, ha, we've got him. He stepped into our trap.

"Faithful, look here. You told us that this first bird is a man sissytautau and this other one, the one you brought us yesterday, is a meri. But, really, Faithful, how can you tell us that this brown bird is also a sissytautau?" And at that moment we realize that we are not the only players in this game.

"This *liklik* bird is a *pikinini.*" In Melanesian pidgin, pikinini is a youngster, a juvenile, and Faithful is the clear winner. For the rest of our stay, regardless of how hard we try to trip him up he is always right. He never gives us two names for the same species, and he always knows which are males or females.

While Faithful has many impressive skills, he apparently has no power over the weather, which has settled into the monotony of daily showers. It is clear that if we are going to stay in this camp for some weeks—and because of the threats we have received, we have little choice—some home improvements must be made. More ferns for the walls and roof are a given. Gusting winds are now accompanying the rain storms and we are getting wet. We also need to do some serious landscaping. Trenches must be dug to channel away rain water, and saplings must be cut and laid down for a walkway between our huts. While our huts are only about ten feet apart, the path is a muddy quagmire from the daily rains. And Frank needs more ferns for his sleeping platform. I don't use a sleeping platform of sticks and ferns. Never have, never will. I always sleep on a cot. Nylon pad, aluminum frame, folds up into a tube about three-feet long, slips into its own waterproof bag, and lashes to my pack. Wherever I go, my cot goes. This is part of a lesson that I learned very early in my fieldwork career. Namely that there are certain comforts that cannot be compromised, and one of these is a good night's sleep. I am not comfortable on a bed of saplings and ferns. Neither is Andy who, a year earlier, took my sage advice and bought a cot just like mine. Frank, however, groans, moans, and plays the role of the macho martyr on his bed of sticks, but that's an I-told-you-so story best left untold.

As long as I am confessing my desire for comfort, I must also confess that I enjoy the comfort of a well-designed commode. Now, I realize that there are some, perhaps many, who will question the need for a luxurious loo. To you, who measure a wilderness experience by the lack of a cold brew, the purchase of new hiking boots, or by the get-away-from-it-all experience of a candlelit bungalow, let me say that life can get wilder. To the biologist who labors for weeks and months in the outback, where brew is never cold, new boots hurt, and candlelight is not romantic, just dim, these little nuisances are simply little nuisances. What I really miss during extended periods in the field is neither ice in my drink nor a newspaper at my door, but the opportunity to retreat at least once each day to the comfort of a well-constructed john. Consequently, realizing that we are staying put for much longer than we had planned, and that the daily experience of rocking on my heels is already getting old and uncomfortable, it is time to resurrect a long-term research project in the design and development of the perfect toilet.

"Faithful, we need to build a toilet." Actually, what I said, in my best Melanesian pidgin was "*Yumi mus wokim haus pekpek.*" He gave me a very surprised look, which I took to mean that the backside of trees had always been good enough for him, why not me. I showed Faithful, he of little faith, the ultimate in pekpek pit design. First is placement: the key to success. A few days in-country gave me a fairly decent idea of prevailing winds (important to know), and these days in-country gave me all the stimulus I needed to get off my tired, old heels. Site selection is crucial—it must be close enough for convenience, but distant enough for hygiene. Selecting and discarding several sites, the choice was made to place the pit on a bit of a knoll, where the visitor could bask in the sunny warmth of a tropical day, and yet be within middle-of-the-night stumbling range if so moved. A mere thirty yards downwind from our sleeping hut seemed ideal. A trail was hacked, roots and stems removed, and we were ready for construction. The design, begun in Mozambique some thirty years ago, and continually refined with time, was sketched out and Faithful was appointed construction foreman, organizing the camp followers to scour the bush for more saplings and ferns. First the pit was dug, and then the walls were built—the back wall a bit higher than the sides to keep the prevailing breezes off our necks. Every decent loo must have a swinging door, with a latch, of course. The door must not be too high, for fear that some rare bird would fly by, heard but not seen, hidden from view by a door too tall. Once the structure was built, the accoutrements were added: a footmat of small saplings (an essential addition in a land so wet), a short stick lashed at a slightly upright angle to hold the paper (a source of amazement to Faithful and his Daga kinsmen), a rack to hold a copy or two of our favorite magazine, a roof of freshly cut ferns to catch those pesky rain drops, and finally a seat made with the greatest of care from specially selected saplings of only the smoothest texture. No rough corners would be tolerated. It goes without saying that the seat must be level.

A full day of diligent construction and the project is completed. This monument to comfort and pleasure is wonderful. The best I can ever remember. Rank has privilege, and as senior author and lead architect, I have first go, then good friend Andy, and finally Frank, our team leader who receives no special toilet privileges due to more than a bit of cynicism over this project. Dinner-time conversation is abuzz with compliments on the pit. As foreman of the construction crew, Faithful is lavished with praise and receives an extra ration of boiled sweet potato *and* first choice of stew meat—he chooses the tiny rear legs of a Ringed-tailed opossum (a wise choice, we agree). Not surprisingly, we all make an extra excuse for one last sit at the pit before bed.

Days become confused, running together with eighteen-hour work routines, constant rain, and increasing fatigue. However, step by step, we begin to build a collection of rare and not so rare captures, and begin to think that this might be a successful expedition after all. Since every bird and mammal has a local name, and Faithful knows these names, we are able to put out a request list of species to the locals and they respond. What is even more amazing is that once we have bought enough of a particular species, and we tell a few hunters this, that species is no longer brought into camp. It is almost as if this information is spread by telephone or telegraph, neither of which exists here. Our trap lines are producing predictable and repetitive results so we scale back on our own efforts and double our energy to buy specimens from local hunters. Each afternoon hunters come into camp to sell us some of the rarest of eastern New Guinea's mammals, including several species that haven't been collected for more than forty years, and may only be represented by fewer than eight to ten specimens in all of the world's museums. Locally abundant, but seemingly restricted to these nearly unvisited mountains, the documentation of these species, and the information that we are able to record about their habits, exceeds our wildest hopes.

Late to bed, early to rise, and sleep comes quickly when so few hours remain before dawn. This night, however, sleep is interrupted. It's raining, as usual, but tonight the wind is blowing, and we are frightened. Wind in the tropical rain forests of the world is the most dangerous of dangers. Forget the beasts, forget the people (unless you happen to encounter the villagers living down slope from our camp), but worry about the wind. Rain forest trees, by design, are top heavy. To be well-rooted is to drown in hundreds of inches of rainfall each year. Put your roots a hundred feet up, on the branches, and cover them with moss and that is how to survive the rains. However, one worries with reason when the winds blow, because shallow roots and heavy, moss-laden tops make for falling trees. In fact, throughout the day, in all directions, we have heard the crashing and cracking of falling trees. One fell between Andy and me as we were setting out traps. It came down so fast, I hardly had time to step back behind another tree for protection. When one of our hunters was setting snares up the trail from camp, he was nearly crushed by a tree that broke in half and crashed down with no warning. We heard the crash, and then we heard him begin to yell continuously as he ran back to camp. We had been trying to persuade him to be our camp cook, because he really wasn't very skilled at snare setting, and with this close call, he agreed to a job change.

It's one thing to hear falling trees during the daylight when you can see what is happening. It is quite another to hear these sounds at night, while laying bundled up in a sleeping bag. Everything is bigger at night. Senses are tuned finer, and rational thoughts become irrational and confused when mixed with sleep. And this is what we hear, at about three in the morning: crack, crack. Slowly at first. Maybe it's my imagination. I was asleep. Did I really hear a crack, crack? Frank is snoring, as usual. No help there. Andy isn't, and his utterance of the Lord's name leads me to believe that, yes, I did hear a crack, crack. Listen some more. Hear my own heart beat. Blood rushing through my ear drums, sounds almost too loud to be real. There it is again, that crack, crack, and this time there is no mistaking the sound. A tree is falling. Very likely it is the tree that is rooted next to our hut. The one that leans out and should have been felled before now. I say very likely this tree because I sense that I can feel the roots pulling up under the hut. This giant emergent, 100 feet high, soon to be 100 feet long, is coming down in some unknown direction. Sleep induced confusion. What to do? Rolling off my cot, I dive for the safety of our flimsy worktable, thinking, somehow, that this eight-foot length of vine-tied saplings will save me from the fall of a forest giant. Andy jumps up on hands and knees and his cot springs vertical like a surfboard gone ballistic. Rolling himself off in record time, he joins me in this not-so-safe haven. And Frank snores on. Crack. Crack. Crack. As fast as cracks can sound, the tree is falling. The most frightening sound in my memory. I just know that my time is up. I'm going to die, this very night, on top of a ridge with no name, in a country that epitomizes primitive. This is to be my last breathing place. Why did I ever agree to the expedition?

Time slows during stress. People say that they can see their life play out just before something really scary happens. I've witnessed this. Most people have. This time my life doesn't do any flashing but I decide, in the few seconds that I have left on earth, that this might be a very good time to pray. Now, I must say that praying is not something I do often. In fact, since I was very young and my Sunday School teacher told me that I must talk to God, I probably haven't tried more than a few times, and I really don't know if I was ever heard. Believing, however, that the direct communication line to heaven is open, and I still have one or two seconds before crush time, I'll place the call anyway. In the few heartbeats of time that it takes for the tree to crash to the ground, I tell God that if I live out this ordeal, I will be a good guy, I won't begin barroom talk with his name, and I might even start going to church. Fright becomes really strong when you know that you are just about one second from death, and at the last instance, when the crack, crack seemed to turn into a roar of limbs snapping and breaking, I even threw in a pledge to believe in miracles if I am spared this horrible, too-soon-to-happen crushing.

CRASH! Black type, all caps, on white paper. Writing the word "crash" cannot do this sound justice. It cannot describe the sensation of roots being pulled up through the dirt and debris of the forest floor, nor can it describe the feel of the ground shaking when a tree weighing tons crashes down. This is violence at its best. Where did it land? Who did it kill? I hear Faithful shouting something in a language I don't understand. Instincts honed from years of living in the bush propelled him far up the trail at the first sound of cracking. Faithful is shouting out to his men, and they answer. They're okay. Andy's okay. I know this because his eyes stare at mine, but no sound has left his lips. And, Frank? He's okay. In fact, he's asleep. The only word he says, in the midst of chaos and confusion, is "Jesus" uttered mid-snore. What a guy!

Get up, grab a flashlight, look for damage. Smashed wood everywhere. Limbs and leaves clutter our path, and the incredibly intense smell of newly exposed earth fills the air. Where is the main trunk of tree in this tangle of broken limbs and leaves and debris? Was it the really big tree that overhangs the camp? Here it is, twenty yards north of the flimsy table that Andy and I sought out for safe haven. It's too dark to look further, too early to begin a new day, but sleep won't come again this night. No espresso ever had the charge that just surged through our veins. Words come fast and emotions are expressed. Frank, awakened by the shouts of our Daga warriors, and by the excitement that Andy and I share, wonders what in the world? Look, Frank. Look at what fell while you were sleeping.

First light comes about 5:30, and with it a reconnaissance. Damn, that was one big tree. And, yes, it was the tree that leaned out over camp. Look how big it is—too large to wrap our arms around. Look how close it came to our hut. We're one lucky group of blokes. By some miracle, or stroke of good fortune, the tree fell in the only direction it could and not hit our camp. Any other way and we would be toast, history, never to kiss our wives again. Constipation is not a problem this day. A trip to the pit is in store. And then we realized that the tree top points directly towards our pekpek house. We thread our way through the piles of debris and discover that the pit is gone, smashed, never more.

We are spared to live another day, and the grand master of all pekpek pits is history. It will be rebuilt, of course, but without the ritual and excitement of the last one. Somehow our hearts just aren't into creating another epitome of design and decorum. We settle into the routine of fieldwork with eighteen-hour days and no time off for good behavior. Days stretch into weeks, and when the time comes to end this expedition, it is in anticipation of a high quality, intense celebration back in Port Moresby, with a hot shower, followed by a hot bath, and copious amounts of beer and broiled lobster at the best restaurant in town. Our mission is a success, the plane returns to fly us out, and we will escape from this strange land alive. You just can't ask for much more than this.

As we are loading the last of our gear into the plane, preparing to leave the valley, Faithful becomes very quiet, and seemingly depressed. He tells us that he is very sad to remain in his small world of a few valleys and mountain ridges. He wants us to take him to the United States where he can have a future. What he does not realize, and we don't know how to tell him, is that his brightest future is in the land of his clan where he is so very successful with a wife, family, a productive garden plot, extraordinary hunting skills, and a very good knowledge of folk medicine. Outside of his valley his chances for success are doubtful, and he could possibly end up like many other Papua New Guinea nationals who migrate into the city—culture lost, alcohol dependent, and sadly lacking the basic living skills for an urban setting. I teach at the University of California, where the students are some of the best and the brightest. Sadly, few possess Faithful's wisdom, skills, hunger for knowledge, and his amazing ability to grasp new information. What a wonderful pupil he would be. And what a terrible burden our society would place on him as he slowly lost his culture while searching for his future.

On the flight out the jungle looks so small. Trees that stand one hundred feet tall appear like toy models below. The Tavenai River, running wild and unchecked, winds its way south to the Coral Sea. Had we not met our plane, we would be on foot, following that raging, brown river, wondering how we could possibly cross and recross its currents until we reached the coast. Bush planes crash all too frequently, but today I will take my chances on any plane that will keep me from a walk out of the mountains. This is reflection time, and I think about a statement a colleague once made.

"Fieldwork," he said, "is an adventure that is filled with anxiety and despair, the routine broken only occasionally by moments of sheer exhilaration." How true. I told this to a friend who had never been on an expedition. His field experience was limited to weekend backpacks and tent camping trips with his family. He couldn't understand the part about despair. But if you have ever found yourself alone, in some remote land few visit, your body bone tired, and your mind filled with thoughts of friends and family back home, you'll understand.

You may even understand why I say that the greatest moment of an expedition, aside from those occasional periods of sheer ecstasy, is reflection of thought, when all is done, and that first cold beer is in hand.

I'm back home now, reacquainted with my family, tending to the routine of the city dweller. The first light of a new day is dawning, and it is quiet. This is the time when thoughts come easy, and my thoughts turn to the night the tree fell. I often wonder if my prayers reached a spiritual set of ears? We'll never know, just like we'll never know if a lone tree crashing in a forest makes a sound. My bet is that the tree does make a sound. The one that almost killed us sure did. I think a lot about my very hurried prayers, and the pledge I made not only to believe in miracles, but to go to church. I'm sorry to report that I have been a bit lax on that front. Promises come easy, especially when they are motivated by sheer terror. But somehow the lure of a Sunday paper and a double cappuccino probes stronger than a trip to the pew. Nevertheless, I think about my prayer and my pledge, and I wonder if living by the golden rule is enough sincerity, or if I'm due for some good, old-fashioned, New Guinea payback the next time danger falls near.

Innocents Abroad

by Robin Dunbar

FIELDWORK, ONE IMAGINES, is the experience of a lifetime. Close to nature, absorbing the ambiance of a real world where even the dust is clean and crystal clear mountain streams sparkle as they tumble out over precipices of infinite depth. Well, yes, there is all that. But, the experiences of a lifetime turn out to have as much to do with learning about human nature as anything else.

I suppose my introduction to the vagaries of life in the field began with a haggle of vultures battling over the carcass of a long-dead donkey one day in 1971. My wife and I had just arrived in Ethiopia, full of excitement and anticipation, to study gelada baboons in the Simen Mountains, an area of breathtaking beauty perched on the mile-high escarpment of the western rim of the Great African Rift Valley in northern Ethiopia.

But all that was to come. Our first few months had been spent cooling our heels in offices in the capital, Addis Ababa, collecting permits and letters of introduction that were necessary if we were to venture so far into the remote hinterland of this ancient country. It was Sunday, and every office in town was closed. So we decided to take a drive out to see the countryside and, we hoped, some *wildlife.* About thirty miles down the road, we spotted the heaving throng of feathers not far from the roadside. We braked hard and reached for our cameras.

We had been clicking away merrily for perhaps ten minutes when we suddenly found ourselves surrounded by the Ethiopian army. We were under arrest.

"Pardon?"

"Under arrest."

"But, whatever for?"

"For taking photographs of a military installation."

I pointed slightly uncertainly at the donkey carcass.

"No, over there . . . a military installation."

Turning to follow the direction of the outstretched arm, we noticed for the first time that the sun was glinting off half a dozen low corrugated iron roofs nestled within a barbed wire enclosure at what seemed like forever in the distance. They could have been anything—a farm, a dispensary, a road construction depot. Everyone puts barbed wire around their compounds in Ethiopia.

Eventually, after much argument and a great deal of disbelief on both sides—as to whether the camp would have been visible in the background of a close-up of vultures taken through a two hundred millimeter lens, and whether anyone could seriously pretend to be taking pictures of *vultures*—we got our cameras back and were released. But our film was to remain in military custody forever. At least, two rolls of film. A third camera had not been noticed, but, unluckily for my budding career in espionage, the film shows only . . . vultures.

~

By comparison with the officers of government, relationships with the people of the countryside are positively delightful. But we found it best to deal with people on their own terms.

Nothing gives the highland Ethiopians, for example, more pleasure than a long and involved battle to extract something from someone else. It's not so much the winning that's important, as the taking part; and the ultimate prize goes to whoever has deployed the most ingenious strategies in doing so. Litigation is an end in itself—something to be savored in the long nights around the flickering glow of the camp fire. By the same token, a man's property is less his possession than something he might, quite legitimately, be relieved of. Inevitably, perhaps, foreign visitors are at special risk of leaving with less than they brought with them.

We learned this the hard way very early on. We had been warned that one of the stunts often played on unwary foreigners by petrol pump (gas station) attendants was not to zero the pump's register before putting petrol into your tank. You must always check that the dials are set to zero before letting the pump nozzle anywhere near your tank.

I had been meticulous in following this advice for months. Then, one day we drove into a petrol station on the outskirts of town. I unlocked the petrol cap, and said, no doubt just a shade too nonchalantly: "Fill it up, and check the tire pressures."

"Yes sir!" said one attendant as he reached down for the nozzle.

"And which tire, sir? This one?" said another, as he beckoned me round to the far side of the Land Rover.

I knew, in that split second as I half-turned to follow him, that I had blown it. The petrol pump was running and, by a miracle that in another time and place would have merited a canonization, we managed to fit five more gallons of petrol into our tank than it was technically capable of holding. It was a seamless piece of artistry, played out with the practiced skill of the Broadway actor. Thoroughly out-classed, the amateur could only gracefully acknowledge the professional.

My long-suffering graduate advisor, the eminent British primatologist John Crook (who had himself worked on the Simen gelada some years earlier), also had cause to rue his enthusiasm for Ethiopia. He had come on a supervisory visit to see what his students were up to. At that time, the Simen Mountains National Park was accessible only on horseback—a journey that took the better part of a day from the roadside town of Debarek 350 miles north of Addis Ababa.

John spent Christmas 1971 with us. When the time came for him to return to Britain, his bags were duly loaded onto mules and we all set off for Debarek to take him to the airport in Gondar, about fifty miles to the south.

It proved to be an eventful trip. First, one of the mules carrying a crate of empty beer bottles decided to divest itself of its genuinely awkward load. As the load slipped sideways, a projecting corner dug deep into the poor animal's side. It went berserk. Bucking and kicking in pain, it raced madly up and down the line of horses and mules, beer bottles tumbling in gentle arcs through the air. Needless to say, the other horses began to bolt in order to escape the flailing hooves and flying bottles. More loads began to slip. Soon, the horses were dispersed as far as the eye could see. Backpacks and suitcases were scattered among the bushes. It took the better part of an hour to round up the strays, collect all bags and the beer bottles (miraculously, not a single one had broken) and reload.

A little further on, the muleteers decided that, after all the palaver, they needed to stop for refreshments outside a small cluster of huts on the path. Such places function both as homesteads and, by tradition, as drive-ins for travelers on the road. The lady of the house duly brought out *talla* (a local beer fermented from grass seed—close to Mississippi River mud in consistency and taste, but after a glass or two you can walk all day).

As we sat sipping the brew, the horses began to drift off in search of grass. A few made for the wattle compound that surrounded the huts, doubtless hoping to find piles of loose barley. No one took much notice. Our attention was eventually attracted by the barking of the household's dogs, irritated at finding strange horses in their territory. Conscious of the ease with which things can disappear given half a chance, I was anxious to retrieve the offending horses and made for the compound.

The muleteers' warning that the dogs in this particular compound were unusually vicious caught me just as I was about to enter. One glance through the open gate at several sets of bared teeth was enough to convince me. Besides, as the muleteers hastened to point out, the horses inside the compound were unlikely to stray further. I remember noticing idly as I turned away that John's suitcase was perched on the top of one of the horses in the compound. It looked in very bad shape, having taken quite a battering from being kicked by a panicking mule during the melee a little earlier. Roped tightly onto the horse's back, it was evidently under considerable strain, with one corner pulled partially back so that the edges no longer quite met.

Refreshed, we eventually regrouped, checked the horses' loads and set off once more. That evening, at the hotel in Gondar, John arrived for dinner somewhat crestfallen. Did anyone know whether his tweed jacket had been packed in someone else's bag? A picture of a horse with a battered suitcase on its back standing in a compound full of barking dogs edged its way into my consciousness. The jacket had gone the way of all things. Lacking another jacket, John returned to England in mid-winter wearing just short sleeves.

Needless to say, no one at the compound knew anything about it when we called back a few days later. What adds to the piquancy of this story is the fact that John had also paid for all the beer himself.

~

Exactly how things got lost, or where they subsequently end up, usually remains one of the great mysteries of the universe. But just occasionally the wheel comes full circle. One such story runs like this:

In 1972, a friend of ours was working on an anthropological project on the edge of the Simen Mountains not far from where we were studying the gelada. Part of his work involved monitoring a small farm at a village called Zarema down in the lowlands at the foot of the escarpment. Zarema was (and presumably still is) a tiny two-bar village on the main trunk road north from Addis Ababa to Asmara. It was far too small to feature on any but the most detailed of maps and was the better part of 100 miles from the nearest town that boasted either a pharmacy or a petrol pump.

One day, the farmer asked our friend if he would like to buy a camera. Intrigued, he asked to see it. It turned out to be a telephoto lens. The farmer had no idea what it really was, but realized that it had some connection with photographs and was thus likely to be of some monetary value to foreigners. He said, somewhat diffidently, that he would be willing to sell this thing for twenty Ethiopian dollars (about ten U.S. dollars at the time). In reality, of course, he had valued it at precisely ten Ethiopian dollars, but the rule of thumb in bargaining is always to begin by doubling the number you first thought of.

Our friend instantly parted with the twenty dollars and promptly bought a camera body to go behind the lens the next time he went to Addis Ababa.

Quite by chance, about six years later, I discovered how the farmer had come to have such an exotic item in his possession so far from "civilization." While at a conference in 1978, I met for the first time an American colleague who, ten years earlier, had been a young Peace Corps volunteer assigned to work in the newly created National Park on the peaks of the Simen Mountains.

One day, while out photographing the wildlife, he inadvertently left his telephoto lens on a rock overlooking the escarpment face. Returning less than an hour later to where he thought he had left it, he was mortified to discover that it was nowhere to be seen. Nor was there anyone in sight for miles around. Entreaties to the local villagers and the park rangers to keep their eyes open for the missing lens yielded nothing at all in the months that followed. It had simply vanished.

Four years later, it turned up in the hands of a peasant farmer 6,000 feet further down the mountain, across trackless terrain that would have taken at least three days to travel on foot. Traded for a bag of barley here and a few dollars there, it had passed from hand to increasingly puzzled hand until it finally found its market value at twenty Ethiopian dollars in the hands of a young anthropologist.

~

We returned to the Simen in 1974 for a second stint of fieldwork, this time determined not to lose any of our possessions. But the problem we faced was how to accomplish this. One thing was clear: For the litigious Ethiopians, nothing (except perhaps for pulling a fast one over a friend) could be sweeter than outwitting a longstanding rival. Clearly, the best option was to enter into the spirit of the game with equal zest.

After a good deal of reflection, we concluded that much of the problem was created by ostentatious display. Too often, foreigners fall prey to the misguided view that their status in society is a function of their apparent wealth. In one sense, of course, conspicuous wealth does confer some semblance of status, but it often does so in a wholly alienating way, setting the individual apart from the community, especially when that community lives as close to perpetual starvation as many do in the Third World. It creates a kind of Robin Hood effect: The rich exist for the purposes of providing the poor with access to things that they wouldn't normally be able to afford.

Our solution was to wear only old clothes and to keep our cameras and other equipment out of sight. Our other strategy was to buy the biggest padlock we could find and fix it to the front flaps of our tent. It appeared to work; people seemed to pity us for our poverty, at least compared to other foreigners who came to the area. More significantly, in a year under canvas, we lost nothing at all. So far as we know, we were the only foreigners who spent any time in the Simen who lost absolutely nothing.

After theft, the second major difficulty researchers face is visits from their compatriots. Remote as we were in the Simen Mountains, we had our fair share of unexpected visitors. It all began with a stray couple hitchhiking their way from Cairo to Kenya down the Nile. They chanced to hear about the Simen and turned aside from their route to see it. They found us ensconced on a remote peak, delighted to see our first western visitors since arriving there ourselves a few months earlier.

We fed and entertained them, gave them a spare tent to sleep in, and gladly took them out to introduce them to the monkeys we were studying. Needless to say, they were fascinated and overawed.

Then, over the course of the next few months, we began to see more and more people. All of them hitchhiking through Africa. All fascinated to see our monkeys, to travel through the Simen on horseback or on foot. The trickle turned into a stream, the stream into a torrent. Soon, we were running what amounted to a free hotel. We began to run out of food—a definite problem when replenishing stocks meant a three-day round trip at the very minimum.

It also began to cost us a good deal of precious time. Evenings that should have been spent writing up notes were spent entertaining. Where we normally would have gone to bed at nine in the evening so as to be up at five in the morning and out with the monkeys as dawn broke over the mountain's rim, our guests would prolong the evening's chatter into the small hours and another working day would be lost.

Then came some real horror stories. There was the group we discovered chasing our study animals for all they were worth one morning in order to get good action shots of a herd of 300 monkeys pouring over the cliffs for safety lower down on the escarpment face, just like we had told them the night before. There was the group who had arrived while an eminent primatologist was on a working visit with us. He kindly lent them his spare mountain tent when it transpired that they were going to sleep in the open—at an altitude where dawn temperatures were invariably sub-zero. One of them was horridly sick during the night, but they were most put out the next day when asked if they would mind washing the tent out before they left.

It was only a few months later that the reason for this sudden deluge of visitors became clear to us. We heard it from the health and nutrition project working in Debarek down on the main road. They too began to find the number of visitors increasing.

Indeed, their final straw came when one group arrived just as the entire project was about to leave for an extended trip to the south. Nonplused, our colleagues extended their unexpected visitors a welcome, offered to let them stay in the house, and set off. They returned two weeks later to find their household staff on the verge of rebellion and the house itself emptied of all supplies. Every beer had been drunk, every precious packet of cookies eaten, the fridge emptied. It seemed that their guests had made themselves thoroughly at home, ordered the household staff about unmercifully, and only finally moved on once they had nothing left to eat or drink. They had bought nothing to replenish the supplies they had consumed, and left the place looking like the aftermath of a cyclone. And all without even a note of thanks.

A few days later, another group of tourists arrived. They were met with stony faces. Much put out at the lack of welcome, they observed somewhat diffidently that they had come all the way back from nearly 100 miles further down the road because they had heard that the Simen was definitely not to be missed; what's more, they had been told that the people who lived there were so welcoming that the whole trip could be done almost for free.

It seemed that we had collectively entered into the folklore of the Cairo-to-Nairobi trippers who, bumping into each other at intervals along the route, would exchange information on good places to see and stay.

Our friends in Debarek placed an instant embargo on visitors to their house, threatened any employee who brought another foreigner to them with instant dismissal, and lost no time in passing the message on to us further up the mountain.

Thereafter, we did our best to avoid bumping into visitors in so far as this was possible. And in those cases when we did encounter them, we operated a simple rule of thumb. If they enthused about the wonders of living in the midst of nature and the marvels of studying animals in the wild, we quietly excused ourselves; but if, after an initial inquiry, they apologized for interrupting what must surely be a busy routine and made to leave, we promptly invited them to dinner. This latter group invariably turned out to be genuinely interested in the wildlife and would never allow us to disrupt our routines unnecessarily or let us feed them wholly at our own expense.

By our second trip in 1974, we thought we had cracked this particular problem. But our calculators failed to take into account the advance of civilization. The World Health Organization was in the middle of its ultimately successful campaign to eradicate smallpox. Ethiopia was one of the handful of countries in the world where the disease was still a serious endemic problem, so much of their effort was directed to vaccinating the dispersed rural population in the countryside. It was no small problem, for the rigors of the mountain terrain had preserved Christian Ethiopia's independence through the centuries while the rest of Africa fell victim to the colonizing powers of Islamic Arabs and Christian Europeans.

With endless funds and the power of modern technology at its disposal, WHO was able to triumph where the very best adventurers and soldiers had previously failed. It simply used helicopters to ferry its vaccination teams around the mountains. For nearly a month, we suffered from the daily buzzing of bored helicopter crews joy-riding up and down the escarpments. Morning after morning, our study animals were sent helter-skelter over the cliff face, tumbling recklessly down giddily inclined slopes in search of safety as a contour-hugging helicopter burst suddenly onto them at an altitude of 100 feet or less without warning. Even the rare Walia ibex, of whom a mere 150 remained alive and who were the entire *raison d'être* for the Simen Park's existence, would race headlong down the cliffs, their massive horns threatening to overbalance them at any moment. Several helicopters came to grief from irate villagers who stoned or fired rifles at these uninvited intrusions into their lives. One came close to a similar fate when it landed right in the middle of our gelada. I was threatened with arrest (again!) for trying to stone it myself, but they did stop flying over our study area.

The romance of fieldwork may come from isolation in an exotic environment, but the technology and congestion of modern societies hover inexorably behind one's shoulder. There is, it seems, no escaping its intrusive presence. It's a small world, and its remotest corners are far from quiet retreats.

Back Seat Observers

by David Bygott and Jeannette Hanby

THE COCKPIT OF the Cessna-182 was about as small as the inside of a Volkswagen Beetle; its single propeller whirred invisibly in front of us. In this fragile cage, traveling at 100 mph, 300 feet above the green woodlands of Western Tanzania, four very different people would be working together during the coming week. We had been commissioned by the Bureau of Resource Assessment to survey the large animal populations and vegetation of the Ugalla River basin and the Lake Rukwa area. These were remote wilderness areas used mainly by hunters and poachers. We would be seeing them from the air only but had heard lots about their wild beauty and richness of wildlife. The report of the survey was a slender pamphlet, with neat computer-generated maps and concise methodology. It did not include any of what follows.

David and I were lion researchers at the Serengeti Research Institute. During the past three years we had also become experienced as aerial observers, partly through a love of adventure and partly by default, as we seemed immune to airsickness. So when the chance came to survey a huge new slice of the country, with companions we knew well, David jumped at it. He loved flying and he was good at spotting animals. I was less eager. My eyes weren't very keen; to do a good job I would have to try extra hard, scan, peer, double-check, and guess. However,

David wanted to go so I accepted the prospect of several days of physical and mental stress. I knew they would need me.

The logistics were planned by Bodgers, an ecologist at our local university. For many years a pilot-biologist with the Game Department, he knew the survey area and had useful local contacts. But he was only free for a week, and his tight schedule made no allowance for breakdowns or bad weather. We would be refueling at several small "bush strips" to which aviation fuel would have to be brought by road. The Bureau provided a Land Rover and three staff members for this purpose.

The aircraft was provided and piloted by Snorts, an ecological consultant from Nairobi. We knew him to be a cool and competent pilot. At this time, the border between Kenya and Tanzania was closed, and it was no easy matter for Snorts to get permission to enter Tanzania, let alone land at the Serengeti Research airstrip near our house, so we drove to meet him in Mwanza. This was an adventure in itself: an epic two-day journey through the flooded western part of the Serengeti and along the pot-holed main road that skirted Lake Victoria. We left our car with friends and met Snorts at the airport.

You can tell a Cessna-182 from afar; it makes a kind of whistling sound above the engine's drone. You can tell Snorts from a long way off, too, but you can't tell him much, as the Mwanza customs officer discovered. He towers over you like a swarthy rugby player in shorts, aims his nose down at you, and blasts you with the Queen's English. Kiswahili just washes off him like water from a duck; he blandly ignores it.

He greeted me with his usual joke: hugging empty air at his chest height, just over my head, and then looking down in amazement as if I had shrunk suddenly. Very funny.

"Had a bit of a bother back there," said Snorts. "Chap tried to tell me I hadn't got the right clearance. I bloody well have! I say, you're not bringing all *that* in my aircraft, are you?" he exclaimed, looking horrified at my bulky survival basket.

I knew well how trips could trip one up, stranded anywhere. A born Girl Scout with a mind for imagining the worst possible scenarios, I had come prepared, not only for the census trip but also for our subsequent safari to Gombe National Park. I had brought along a compact kit in case of minor and major disasters: medicine, food, water, ground cloths, bedding, and extra *kangas* (the colorful cotton cloths that East African women wear) to use as wraps or mosquito nets, sheets, or towels.

The fact that the plane had no survival kit of its own was a bad omen. I was "allowed" to take my kit. I looked up at Snorts. "You'll be glad I brought this kit, later on!" I said, hopefully sounding tough and self-assured. But there was also the issue of claustrophobia. I have never liked being in enclosed spaces for long, especially when crammed in among such overbearing characters: three bright, experienced, opinionated, excruciatingly funny, but gallant Englishmen. One short, fubsy woman. Sigh.

We loaded the plane and took off, heading south, for Tabora where we were to base ourselves for the first few days and rendezvous with Bodgers.

Tabora is a large town set amid flat wooded country, and has a large airstrip where even jets can land. A big red fire truck stood at the ready, like a hyena awaiting an opportune kill. Bodgers was there on the tarmac to greet us, hands on hips, beard and belly thrust out, brusque, bouncy, bristly, bombastic, but ultimately benevolent.

"Good-oh, you all made it! How's Nairobi? . . . and the lions? . . . Splendid. Well, let's go back to the hotel for lunch and then we can go over the plans."

The Tabora Railway Hotel was a whitewashed, pillared relic of the Raj, and in most of its rooms a fair-sized giraffe could have stood, without stooping like the aged waiters. In the dining room, ancient ceiling fans hung motionless in the upper gloom. Tablecloths and napkins were threadbare but scrupulously clean. Soup and bread were brought to the table.

Bodgers wrinkled his nose. "Don't know what's in it—smells like weevils though!" (These small beetles impart a very characteristic musty smell and bitter taste to old stored flour and similar foods.)

"Weevil see what the bread's like," David suggested. In the company of his fellow Brits, his sly sense of humor came out, and they vied to produce the worst puns.

"All will be weevealed at the first bite!" I said, swigging water to wash away the taste.

"Well, whatever happens, weevil overcome!" declared Bodgers.

"Stop it, you lot!" groaned Snorts, "or there'll be a weevilution!"

"Why?" said David, "You can't just be a wise monkey and see no weevil, hear no weevil, and eat no weevil. . . ."

Bodgers rolled his eyes and intoned: "O Lord, deliver us from weevil!"

"Enough weevolting jokes," I gasped, "and pass more water!"

I had to drink water, lots of it. To my horror I had developed an infection and I had none of the correct medicine and would not find any on this trip. There was a simple remedy however, which was to clear the infection by drinking *lots* of water every day. So far, the pain was mild because I'd begun drinking immediately. But in a plane? With three, well maybe two, unsympathetic men? As David said, "You can't just park your plane on a cloud and have a pee." I was reminded of a terribly English scientist-pilot we knew, who had drunk too much tea before flying alone to an important meeting. In agony high above the

Serengeti plain, he relieved himself into one of his shoes, the only receptacle he could find. The real problem came when he opened the window and tried to empty the shoe outside—a howling wind blew it all back in and soaked him to the skin. . . . No, water therapy wasn't going to work. I would have to fly dry during the day and drink at night. Yes, I was in for trouble.

After lunch, Bodgers brought out a big map with north-to-south lines ruled on it. These were our flight paths, which would keep us busy for eight hours or more, every day of the coming week. The next few days passed like the endless roll of lacy green woodland unreeling beneath the plane.

Flying was less taxing at first than I'd feared. Our transects took us over vast stretches of sparsely populated country. Snorts was responsible for following the agreed flight lines at the correct height. Seated next to him, Bodgers helped navigate, and recorded on a checksheet various measures of vegetation, topography, and land use. David and I counted wildlife from the rear seats.

Animals were scattered and fun to pick out of the leafy, bushy landscapes: roan antelope, greater kudu, hartebeest, as well as elephant and zebra, giraffe, eland, and buffalo. We talked into our tape recorders, eyes fixed on the strip demarcated by the lines drawn on the window and the streamers attached to the wing-struts. Cameras ready, we photographed any large herd, and estimated its numbers. It was especially hard to count the dense herds of cattle, sheep, and goats (or shoats as we called these undifferentiated mobs) and this would be done later from the photos. We noted beehives and huts and livestock corrals or *bomas.* At night we had to play back our tapes and record the data on checksheets. And I had to drink drink drink! My familiarity with the "facilities" was instant and intense. The discipline of the census routine was demanding, tiring, but endlessly interesting. The work helped me forget my worries and the pain.

Being airborne was fun, and in between we were based at Tabora. We became heartily sick of weevil bread, weevil spaghetti, weevil custard, and above all, weevil jokes! Unpalatable as the food was, the water looked worse, which was bad news for me. Ever had to drink iodine-treated water in abundance? Warm beer existed, but we never seemed to be allowed more than one bottle each at dinner, although the regulars out on the verandah had scores of beers stacked under the little tables where they sat, drinking each dusty day down.

We were relieved to move to our next base. Inyonga was a sleepy little village, and ours was the first aircraft to land on its grass strip in ten months. The airstrip boasted no fire engine, but a man wheeled out a trolley with a fire extinguisher as we landed. By the time we had opened the doors, almost everybody in the village was there to stare at us. We signed the register and walked to the government resthouse, one of those buildings where its simplicity is its beauty; whitewashed rooms clean and bare but for beds, the window just a wooden shutter that opened to a view of scarlet flamboyant trees. On the breezy verandah, we unpacked our lunchboxes from the Tabora Hotel; weevil bread, gray boiled eggs and rubber chicken. The tuna and mangoes from my kit were welcomed.

Mangoes were one of the delights and salvations of the trip. No matter how bad the food anywhere, we could always find the delicious stringy-sweet fruits. David took four from the boys who had brought them down by throwing rocks into the tree by the rest house. He caricatured our faces on them and we had a good laugh at ourselves, a green, queasy-looking crew. The soft afternoon consumed our energy just transcribing data. The support car still had not arrived, so we couldn't refuel and fly again. We were all glad to take a break and amble through the village. It was especially pleasant for David and me, because park rules prohibited us from taking walks in the Serengeti where we lived.

During our walk, the ground crew arrived, having driven over 120 miles from Tabora. Bodgers greeted Deogratias, the student from the University of Dar es Salaam.

"What took you so long?"

"The roads were terrible. We had to go slowly because of the weight of fuel."

"Too bad. We had to cancel our afternoon's flight. What to do? (One of Bodgers' favorite expressions was usually said as one word: "Whatudu?") We'll meet you over at the strip to refuel." Snorts and Bodgers headed that way on foot.

Deogratias glared at us from behind thick-framed glasses. He had been trained in Canada to interpret satellite images, not to deliver avgas, and had hoped to see his country from the air, rather than the ground. He was getting more work and less glory than he deserved.

"Aren't you going too?" he asked us.

"No, we'll stay and look around."

"Look at what? Is it a zoo?" he sneered.

"No, it's like a wonderful botanic garden," I said, trying gracefully to smooth what promised to be a rocky road, "with all kinds of foreign trees. . . ."

"Foreign to whom?" declared Deogratias. Nationalistic hackles were bristling under his trendy denim jacket.

"Well, there are jacarandas from South America, mangoes from India, flamboyant trees from Madagascar. . . . Anyway, let's meet later, okay?"

We did indeed feel like alien tourists, dropped from the skies among people for whom the technology, principles, and aims of an aerial sample count must surely seem absurd. To us, in contrast, our own aerial sampling of rural Tanzania was fascinating at a personal level. Stick a pin in a map and go there, and you will always find treasure!

Almost immediately, an elderly man greeted us, wearing a long white shirt and wrap-around *kikoi* and an embroidered Muslim skullcap. His face was long and melancholy, his nose fine and Arabic.

"Come!" was all he said, as he took me by the hand and led us to his home. A heavy wooden door set in a high wall opened into a narrow courtyard. There was a stone-walled well and a small verandah. Beside the verandah grew a small orange tree, supporting a climbing rose bush. It bore one pink, perfect rose. Our host plucked this, and gave it to me.

"Please. For you," he said gravely, and I happily pushed it into my hair, thorns and all.

He brought out a table and chairs for us, and prepared wonderful, sweet, milky tea fragrant with cardamom.

"I am here for thirty years," he told us. "I came from Muscat, Oman, as a boy, and now"—fingering his white fringe of beard—"I am an old man."

Perched on the edge of his well, he told of days when Europeans used to fly in from Tabora, Arusha, even Nairobi, to hunt animals near the Ugalla River. But none had come in the last four years.

"Everything is going down," he sighed. "The whole system is rotten, from bottom to top. If you want flour, you get problems. Clothes—problems. Sugar—problems. The politicians have no sense. In the old days, there was one white man, the District Commissioner, who traveled the whole district on foot. He was in his office at 8 A.M. and if you had a problem, that man saw to it. Now, they come in cars, drive, drive all the time, they never have time to see what is going on, all they care about is drink and women."

"If it's so bad, why do you stay? Why not go home to Oman?"

"Because I am used to this life. An old man cannot learn new ways. My children are grown and gone to Oman and London. My son in Oman makes much money, and I am well looked after. But I swear to you, if these people here ever start causing me trouble, before the sun sets I shall be on my way!"

I thought of the great variety of people we had met in Tanzania. A few were exotics, transplants from other countries, like this Arab, scattered Germans, British, Poles, Greeks. Even the Tanzanians themselves were a mix of more than 120 different tribes. Few indigenous people seemed to think of themselves as Tanzanians; loyalties were local first, then national, reflecting perhaps the recent settlement of these areas by the people, or the recency of national independence. Yet one of the most striking things about Tanzania was that its diversity was combined with an overall tolerance of different customs and languages. Perhaps this contributed to our feeling of being part of the whole fascinating array. If this was a zoo, it included us.

I watched the Arab's noble face fade into the dusk as David rose from his seat, saying, "Our friends are waiting, we really must go." We thanked him sincerely for the tea, and the flower: a priceless gift. For the first time, he cracked a grin.

"It's only a flower!" he chuckled, and in the courteous African manner he escorted us some distance from his gate to see us safely on our way.

Walking home in the dusk was magical, with fireflies sparkling against the dark masses of the mango trees; embers glowed red inside mud-brick houses, and people on foot or on bicycles murmured greetings as they passed us. There was a warm, soft, friendly feel to the night.

The next morning we had no time for breakfast. We were in the air as soon as it was light enough to see. As the Cessna turned at the end of each long transect I would madly spoon my survival-kit tuna onto crackers and hand them around. I thought how lovely it would be to have some calming delicious tea again at the Arab's courtyard.

We continued all afternoon, as we had lost half of yesterday and were now under pressure. That evening, we landed in Mpanda. Once a booming mining town, its deposits of lead and gemstones had mostly been worked out and now it had a listless, ghost-town atmosphere. Our ground crew greeted us with the news that the government rest house was full, so we found lodging in a "guest house," which had a barrack-like row of dark, poky rooms. Clearly, their creaky beds and crusted sheets were used mainly on a very short-term basis.

Daylight faded. Snorts was getting nervous about a large woman with filed teeth who kept eyeing him speculatively, and there seemed to be no food on the premises. In this part of the world "guest houses" often serve no meals; you eat at a "hotel" which has food but usually no beds. Bodgers suggested we look for one. Just then, the ground crew came back to report on their own dinner. Their faces were long.

"It was terrible," said Deogratias. "We went to the Nawaz Hotel, and they served us cold rice, and the meat was bad. And they took forty-five minutes to warm it up for us!"

"*Pole*, chaps!" Bodgers commiserated, "Whatudu? It doesn't sound too promising for us!" He led us down a long dirt street between mud huts until we came to the Kigoma Hotel. We sat at a peeling laminated table, the only diners present. A couple of gangly youths lounged against the wall next to the menu.

This was a blackboard listing all the delicacies which were available, or had once been available, or might one day be available. It started with the simple fare:

1. Chai (tea)
2. Silesi (slice of bread)
3. Hafkeki (half-cake; any wiser?)

and progressed to the more elaborate edibles:

13. Kuku na Wali (chicken and rice)

This sounded good, but they didn't have it. So the three men settled for:

15. Nyama na Wali (meat and rice)

Fortunately, I wasn't hungry, as I seldom eat at night. When the food came, there was a dish of cold, gray-green greasy rice, and a bowl of tepid oily soup in which floated fragments of black vegetation and small gray gobs of elastic meat—mostly leathery tripes, nubs of gristle, and splinters of bone. Manfully, David closed his eyes and took a few mouthfuls, but soon found that it didn't taste nearly as good as it looked. Snorts quietly put down his spoon and coughed. David, who will eat almost anything, pushed his dish away. Bodgers muttered, "Goat's vulva was never one of my favorites."

The bill was presented to Bodgers. "Do you expect us to *pay* for this?" he roared, as he pushed back his chair and rose to his feet—an impressive figure with his high forehead, bristling beard, and icy eyes. He called for the manager. One of the lounging youths straightened up and simpered. Bodgers puffed out his chest and declaimed theatrically, in fluent Swahili: "Mr. Manager! We are visitors from afar. We have traveled from Europe and America to see your country. We have dined in many different places and eaten all kinds of food. We have eaten in Dar es Salaam, in Mwanza, in Arusha, in Tabora, we have eaten in Kigoma, Dodoma, we have sampled the delicacies of Inyonga and Urambo and innumerable other towns. In some of these places the food was poor, in others it was excellent. And now we have eaten at your fine hotel; it is my pleasure to inform you, sir, that out of all the dinners we have eaten in all these different places, not only in Tanzania but in other countries as well, the meal you have just served wins the prize, sir, the very first prize, over all others, Number One completely, as the *most disgusting filthy garbage that we have ever tasted in our lives!* Good night!"

Bodgers waved some greasy notes in the air then plopped them onto the counter. He should have had a cape to fling around himself like Falstaff as he pompously led us into the night. I was last, as usual, and glanced at the manager as he gathered up the money. The corners of his lips turned up in a resigned smile; he gave a small shrug that said "What to do?" I returned this universal human gesture and followed the others back to the guest house.

We transcribed data in the dark by torchlight, under a mango tree across the street because of the noise at the guest house. When we finally got back to our tiny room, I took one look at the bed and cringed. We turned the mattress over and I pulled out my kangas to use in place of the sheets. We closed our eyes and immediately the bar next door surged into life. Chairs were dragged screeching to their tables. Then came the clank of bottles and the hiss of opening them. The radio crackled and whined at high volume. Voices rose to compete with the radio. Doors slammed, feet clattered up and down the courtyard. Soon I had to join the throngs visiting the "facilities." It was hours before it became quiet enough to hear the armies of mosquitoes again.

The dawn was gray and our eyes were red. Determined never to spend another night in Mpanda, we told the ground crew to press on to Sumbawanga, on a high plateau overlooking Lake Tanganyika. We flew transects all morning, but met heavy rain and had to grope back to Mpanda to land. We sat miserably watching rain drip off the roof of the airstrip guard's hut, and decided that the best thing to do was to go straight to Sumbawanga, doing one transect on the way.

Crossing the green floodplains of Katavi, we saw the largest herds of animals so far: topi, elephant, and zebra. Flocks of white egrets wheeled above immense stampedes of black buffalo. Then driving rain set in, and we found Sumbawanga completely socked in with clouds.

"I'm not landing there!" said Snorts. "Any ideas, Bodgers?"

"Hmm, whatudu?—How about Muze? That's down there between this escarpment and Lake Rukwa. Locust Control has a strip there."

"Good show! Let's land and take a break, see if Sumbawanga clears up later."

The lakeshore was a huge grassy plain which used to be a major breeding ground of red locusts. Control planes had used Muze as a base for spray sorties but no control had been necessary for the past ten years. The control organization and the base were practically defunct. For us, the place would be a welcome haven for what we thought would be a short stop.

We circled once to check the airstrip. Surrounded by trees, it looked well-maintained: no aardvark pits, no termite mounds. A ragged-looking man limped out and looked up at us, then ducked into a round hut. He was back at the strip by the time we landed, dressed in his khaki uniform, standing at attention, holding a bucket of water! He was the local locust officer, and as we emerged he hobbled over to greet Bodgers, whom he knew. Crowds of villagers, mostly children, appeared from the maize fields surrounding the strip. Many children looked unhealthy, with potbellies, runny noses, dull eyes. Snorts and Bodgers signed us in while we gazed at the striking setting of this little village; and the villagers in turn enjoyed the unusual entertainment that had dropped into their midst. As we waited, there was a sudden commotion. A neat circular space appeared among the children, a small snake at its center. An old man with boots went and stomped the snake to pulp, and the crowd laughed uproariously. But some dangers are less easily dealt with.

On either side of the hut, stretching almost to the end of the strip were piles of oildrums. At first I thought they were filled with fuel, and wondered how they had survived theft and pillage. But then I noticed that many of the drums were leaking; children were playing round them unconcerned. It couldn't be fuel.

It was much worse. I noted the faded lettering on the drums. In English, a language few people here could understand, it said: *Danger. Dieldrin.* My stomach lurched. These trusting people seemed to be utterly unaware of this malignant poison. My Kiswahili could not cope with trying to explain even if they would have listened; after all, they'd lived with the evil stuff for years and never connected illnesses with it. Pesticides and medicines are all "dawa" in Kiswahili; a cure for a problem. I frantically tried to enlist Bodgers' help as interpreter. He said that of course the people there knew full well what was in the drums and that it was not to be touched. But who would ever move it, where to, and what could one do about rusting and seepage? No one could keep the children away either. "Whatudu?"

Someone brought us mangoes which we ate with gusto, while Snorts watched the weather. It seemed to be clearing up on the plateau.

"Let's give Sumbawanga another try," he said.

After saying our good-byes we scrambled to take off. The escarpment rose sheer for 2,000 feet, with waterfalls streaming over granite rocks amid the shaggy forest. We circled between it and the broad placid lake, slowly climbing. But clouds still hugged the top of the plateau.

"No go, chaps," said Snorts.

"Better get back to Muze, fast! Don't like the look of that rainstorm!" warned Bodgers. Dark clouds were sweeping over the lake towards the escarpment, threatening to cut us off. Snorts wasted no time. Left wing down, rudder hard right, and we screamed earthwards in a sideslip, leaving our stomachs somewhere up near Sumbawanga. As we neared Muze's strip, fat raindrops starred the windshield and we could only see through the side windows. Snorts lined up the plane as best he could, between the trees, lurching in gusts of wind.

"On finals now," shouted Snorts, "but I can't see! Can you read off the radar-alt as I go down?" Calmly, Bodgers called off the heights from the altimeter: "300 feet . . . 200 . . . 150 . . . 70. . . ."

(Trees were hurtling past the wing tips. We were going to die. . . .)

"30 . . . flare out now . . . *Jesus!* . . . about two feet off. . . ."

Forty knuckles were white as we bumped down and finally skidded to a halt a few feet from the tall corn and termite mound at the end of the strip. Snorts killed the engine and slumped, drained, in his seat.

I had new faith in our pilot and copilot, they took on haloes in my joy to be safely on the earth again. Obviously we had to stay the night. The locust control officer said we could sleep in the old locust research building, now used as a clinic. When the rain eased, he led us in procession to the old colonial-style, decaying building on a low hill above the village.

My companions were a man and little boy who carried my survival kit on his head. The father asked me if I knew Kiswahili. I knew just about enough to say, "Not much. We work in Serengeti National Park with the lions where we only speak animal languages. But I am trying to learn Kiswahili." He seemed delighted with my effort and kept saying, "*Mama Simba anajua Kiwanyama na anajitahidi kujifunza Kiswahili*" [Mama Simba knows animal language and is trying to learn Swahili], as he led us to our sleeping quarters. The old laboratory was dank and dark. Bats flickered in and out of holes in the stained ceiling, and their smell lingered even after their accumulated droppings had been swept away. The kind villagers brought mats onto which I spread my formerly ridiculed bedding. Bodgers was in his element, organizing things, dispatching youths to get rice, chicken, and wine. Tired as we all were, the occasion was taking on a festive air.

We sat on benches and waited for results in another large, bare room. Bodgers' boys reported that food was being prepared; he rewarded them with some of my candies. They went to sit on windowsills, giggling and muttering softly about "Mama na Baba Simba" and the "Piloti," and passing around the sweets so everyone could have a suck. A small crowd gathered. The village chairman (a gaunt old ex-policeman from Dar) appeared, and amidst good-humored banter, was given the only chair. Adults and children sat on the floor all around the walls and more little faces peered in the window openings.

A pretty nurse joined us; her name was Sijapata, which means, "I haven't got it yet." (Perhaps her parents had been trying for a son, after a string of daughters?) She and some friends began softly singing a hymn—with that amazing African ability—in perfect tune and close harmony, transcending the rain and the mud and the darkness with pure joy.

"Christ, I wish I could sing like that," said Snorts. "What songs do we all know the words of?"

"Row, row, row your boat?" I suggested.

"Boring!" groaned David. "How about 'Cessna Cessna?'"

Bodgers boomed, "Yes, you chaps sing and I'll translate." So, raggedly but with enthusiasm, to the tune of "Freight Train" we sang David's homemade anthem of animal counters:

Cessna, Cessna, flying so low,
animals flash past my window;
Was that a dikdik or an oribi
My God you nearly hit that tree!

Pilot, please don't think I'm rude,
but could we increase our altitude?
I don't want to seem a bore,
but there are horns sticking through the floor!

Cessna, Cessna, flying so high,
ten thousand feet up in the sky;
Those things down there look like ants,
although they might be el-e-phants

We've been flying ten hours and more,
I can hear my partner snore;
We just counted a herd of sheep,
and the pilot fell asleep. . . .

The laughter that greeted this was probably as much due to our discordant performance as to Bodgers' witty translation. By now, most of the church choir was present, and they treated us to more sublime melodies and harmonies. When the chicken and rice came, it was hot and delicious. We ate with our hands from a large platter and toasted each other with sweet, red

Dodoma wine. A young man joined us at the food platter, very drunk. The other villagers frowned at this breach of manners but he insisted on eating with the *wazungu* (white people). He had some *chibuko* or local millet-mash brew, in a plastic bucket, and passed it around. David politely tried it; he later said it was thick, heavy, and bitter, just like cold vomit. He passed it on, saying it was wonderful but he was too full to do it justice.

Then the old locust officer made a speech, telling the children how Snorts was the pilot and he had been flying us around in the clouds with God, and had brought us safely through the bad weather to land at their village with the help of divine providence. The idea of flying with God was obviously fascinating.

"You know," said Snorts, "it's at times like this that one wishes one knew more Swahili."

"But you've been here for years," exclaimed David. "Are you serious?"

"Well, I know *Jambo . . . Moja* is one, *mbili* is two, but what the hell comes next? . . . Anyhow, most people understand English if one speaks it slowly and loudly enough, you know?"

There are many white people in East Africa who don't know any Kiswahili, but we reckoned Snorts knew a lot more than he let on. After many good nights and sleep wells, we thanked our hosts and trooped off to our bat-room. Soon we were all lined up like sardines on our mats, to share our coverings and the mosquitoes. It was then that we discovered that there are two kinds of bedfellows; those who snore loudly, and those who are kept awake. We knew that our pilot was a snorer, and soon found out that Bodgers was his equal. They made a pride of roaring lions seem quiet.

Even though we hardly slept through the snores of our comrades (who claimed the next day that they never slept at all), we felt refreshed in the morning. We took extra time over a small breakfast, enjoying the cool fresh air as we sat on the steps of the old locust control building, savoring a rich and special experience, looking at the little plane parked ready by the patch of maize and dwarfed by the backdrop of the escarpment to which we were next headed. At last we had to go.

To this day our census trip is captured in my mind by the view as we flew up and off, looking down on those ranks of dieldrin drums along the almost abandoned airstrip symbolizing the ways priorities change as humans face different disasters, be they locusts or diseases or "development." We will remember warmly the upturned faces of the innocent and friendly people, the clusters of huts, the dark mango trees, the wide fields, and the distant lake with plains full of the wildlife we had come to census.

After another day, our fieldwork was finished. Back to Tabora and the crew dispersed. Snorts had agreed to fly us to Kigoma and perhaps join us for a day or two in Gombe National Park, where we were to spend Christmas and the New Year. The country seemed blanketed in rainclouds, and we had to fly low beneath them as we crossed the Malagarasi Swamp. It looked like the most desolate place on earth, an endless sheet of gray water choked by papyrus and reeds. The only solid ground was the causeway of the Central Railway Line, and we followed it through the rain all the way to Lake Tanganyika, listening anxiously to the steady hum of the engine. It was a relief to land on the red earth of Kigoma airstrip.

Snorts vacillated for a long time, as we stood in the wind and drizzle by the plane. Should he come with us, or go home to Nairobi? He shrugged, imitating Bodgers' typical gesture and said, "Whatudu?" We smiled in sympathy. Snorts was tired, the job was finished except for the paperwork, and family ties tugged more strongly than the lure of yet another Tanzanian adventure. He climbed back in the plane and was soon a dot in the sky.

After a taxi ride to Kigoma town, we took a colorful overloaded boat along Lake Tanganyika to the chimpanzee forests of Gombe. Our heads were full of experiences to digest. My worries about flying, "my men," the discomfort, were over. All had been discarded like so many mango seeds and my view was forward, to another scene. David would be reliving forest patrols with his hairy friends. Bodgers was flying back to Dar to meet his wife for Christmas. I wondered what he would be thinking as he crossed the huge country. From on high he would see vast tracts of wild country dotted with cleared settlements, like small burns on a green carpet: a broad sweeping view like his perspective on wildlife, education, and development in Tanzania. Then there was the tired crew on the ground, in the support vehicle racing back to Dar es Salaam, happy to be out of the bush and headed back to the civilization they preferred. We had all glimpsed one another's very different points of view during our brief shared safari.

It was some months before we next saw Snorts in Nairobi.

"How did the analysis go?" we wanted to know.

"Oh, not bad," he said. "Come to the office and I'll give you a copy of the report. Ha, I had to laugh though. They had a big meeting in Tabora and called me down to present the report. So I flew down, said 'here you are, chaps,' opened my briefcase, and found I'd left the bloody thing on my desk here. Damned embarrassing actually. . . . Had to go back and get it, of course, but for all the notice anybody will take of it, I probably needn't have bothered!"

Afterword

by Nigel Barley

FIELDWORK IS A major rite of passage for many of the behavioral sciences. In many departments, no brown knees, no doctorate. It is a guarantee of authenticity and personal seriousness. Like a Japanese *yakuza's* willingness to slice off a finger joint, it shows due deference to superiors and the embracement of suffering to a higher end. Yet it is also a confirmation of the institution of learning as a neutral space from which researchers go out like young lions and drag back raw data to be flensed and processed according to value-free scientific method. And like many framing devices, its power comes largely from its invisibility. For while institutions of learning accumulate endless fieldwork photographs, they generally show "the natives" dancing or herds of animals behaving "naturally." The fieldworker is carefully kept outside the frame. This is not to say that fieldworkers are never pictured in the field, indeed we all have drawers full of "personal" shots of our living quarters and detested colleagues incongruously met up a mountain, the day the police raided us, and the aftermath of the car crash. Come across, years later, they plunge us into nostalgic reverie.

But such snapshots have not, until recently, been seen as an integral part of the process by which knowledge is produced in the first place. Any such snapshots that find their way into working institutional archives have often been weeded out or housed separately or sometimes killed by neglect through lack of conservation. Only lately has it been realized that they tell us rather more about the fieldwork method and its consequences than the "official" pictures, and everywhere such snapshots are now being hastily reintegrated into the total collection.

In papers and monographs, the researcher is traditionally detached and aloof, blessed with an Olympian vision where the world exists in the passive mode. Fieldwork was described to me as a student in terms of a process of progressive purification. One sloughed off the material cares of the West. One merged with the background and was "accepted" by grateful research topics. Interpreters were only hired to be swiftly dispensed with. Never, ever, was one bored or bad tempered or homesick or drunk or even hungry. Love, sex, and diarrhea did not occur even in the footnotes. Pure spirit did not permit of such things.

Humor and publication are similarly segregated in academic life. Piety is, after all, the measure of the worth of one's research and even confessional, post-modern fieldworkers are unrelievedly solemn. Holistic behavioral sciences inevitably require things to fit and click neatly together. They have no room for the incongruous or the downright contradictory so there is no record that Archimedes dared to laugh before he shouted "Eureka!"

It is the strength of this assembly of tales, that they put the fieldworker right back in the frame where he or she belongs. Had such a book been available in my graduate days, perhaps fieldwork itself would have been a much less shattering experience than it turned out to be. These stories carry much of the earthy flavor of the embodied fieldwork experience and its very concrete concerns. They will evince enthusiastic nodding of the head from any previous practitioner. They convey the certainty that, whoever it is that directs the fieldwork process, it is seldom in any simple sense the researcher, who feels more like a ball in a pinball machine than a major player. In being frank about the less comfortable side of the activity, these recollections also go some way towards explaining why fieldwork, against all good sense, is so habit forming; why some people become professional fieldworkers or spend their lives in a constant alternation between "home" and "the field." Between the lines comes the message that, for many, fieldwork is one of the most intense and concentrated experiences of their lives and ordinary existence is bland without it. Moreover without that tattered field diary, dappled with great sweaty thumbprints and crushed insects, the human memory cannot hang on to the really awful moments described here. As one local informant expressed it: "You came here young and went away an old man. Now you've come back young again. Why?"

About the Authors

Nigel Barley

Nigel Barley is Assistant Keeper for West and North Africa in the Ethnography Department of the British Museum. After a Doctorate in Old English magico-medicine, he conducted fieldwork among the Dowayo people of North Cameroon, the Toraja of Indonesia, and urban potters in Morocco. His written works include *Adventures in a Mad Hut, A Plague of Caterpillars, The Duke of Puddle Dock,* and *Smashing Pots: Feats of Clay from Africa.*

Elizabeth L. Bennett

At the time of the trip to the Che-Wong, Elizabeth Bennett was a student at Cambridge University, doing research on the rain forest primates of Peninsular Malaysia for her Ph.D. Since 1984, she has been working in the Malaysian state of Sarawak in northern Borneo. An associate research zoologist of Wildlife Conservation International, New York Zoological Society, and consultant to WWF Malaysia, she works with the Sarawak Forest Department on wildlife research, conservation, training, and management.

This article is dedicated to the memory of Kalang anak Tot, who passed away in 1987.

Richard O. Bierregaard, Jr.

Richard Bierregaard, Jr. grew up in southern New York and was set out on the trail that would eventually lead him to the Amazonian rain forests by Frank Trevor, an inspirational high school biology teacher at the Millbrook School. He received a B.S. from Yale and his Ph.D. from the University of Pennsylvania. He has studied the ecology of birds of prey in the high plains of Montana and on Martha's Vineyard. In 1979 he moved to Manaus, Brazil, where he was the first field director of the Smithsonian's Biological Dynamics of Forest Fragments Project. In the Amazon, he studied the ecology of understory birds and their response to habitat fragmentation. He is now an adjunct member of the Biology Department at the University of North Carolina at Charlotte.

Monique Borgerhoff Mulder

Monique Borgerhoff Mulder first traveled to Africa in 1979, driving with Tim Caro from Algiers to Tanzania. She returned to conduct anthropological research on polygynous marriage with the Kipsigis of Kenya between 1981-1983. After attaining her Ph.D., she started working with the Tanzanian Datoga, the subject of this story. She is now Associate Professor in the Department of Anthropology at the University of California at Davis, and together with her husband, Tim Caro, and their son, has started a new project in southwestern Tanzania, focusing on conservation issues. She is extremely grateful to Momoya Bashgei Merus and Daniela Sieff without whose help the pilgrimage to Gitangda's grave would have been impossible.

David Bygott and Jeannette Hanby

David Bygott and Jeannette Hanby are biologists who had worked separately on chimpanzee aggression and monkey sex before coming to Serengeti together to study the behavioral ecology of lions. It was during the period of their lion work that this adventure took place. Although the story is not about lions, they chose it because so much of a research project is about learning to see and adjust to the country one is in, and the companions one works with. After leaving the Serengeti, the "Hanbygotts" lecture-toured the U.S., then returned to Tanzania to work on conservation education projects. They now have their own company, Kibuyu Partners, and work mainly as writers, artists, and publishers from the remote home they have built near a village in northern Tanzania. There they find the opportunity for adventures every day, instead of just now and then, and a refreshing diversity of points of view.

Tim Caro

Tim Caro is a zoologist at the University of California at Davis whose interests focus on the behavioral ecology and conservation of mammals, especially in the tropics. He has worked on a variety of species: cheetahs, Thomson's gazelles, white-tailed deer, and domestic cats. He has been to many countries in Africa and loves to travel to areas where human impact is minimal, and animals live undisturbed. As a result of his work in Serengeti he wrote a book called *Cheetahs of the Serengeti Plains: Group Living in an Asocial Species* (University of Chicago Press, 1994).

Dorothy L. Cheney

Under the guise of studying the behavior and communication of vervet monkeys, Dorothy L. Cheney managed (together with Robert Seyfarth) to maintain an intermittent residence in the village of Ol Tukai in Amboseli National Park, Kenya, for eleven years. The people of Ol Tukai tolerated their research with polite amusement, and they are grateful to them for their friendship. Cheney currently holds a position in the Department of Biology at the University of Pennsylvania.

Ronald E. Cole

Ronald Cole is a Senior Museum Scientist and Curator of the Museum of Wildlife and Fisheries Biology at the University of California at Davis. Over the past twenty-five years, he has participated in scores of museum research and collecting expeditions to five continents.

Robin Dunbar

Robin Dunbar spent most of his early years in Tanzania. Having earned a B.A. in Philosophy and Psychology at the University of Oxford in 1969, he went on to do a Ph.D. at the University of Bristol under the supervision of John Crook. He and his wife, Patsy, spent two extended periods in Ethiopia during 1971-1972 and 1974-1975 studying gelada baboons, and later carried out a field study of klipspringer antelope in Kenya in 1980-1982. Having held teaching and research positions at Cambridge University, Stockholm University, and University College London, he is now Professor of Psychology at the University of Liverpool in England.

Lisa Halko and Marc Hauser

Lisa Halko was brought up in the eastern United States. She studied acting at the College of Performing Arts of SUNY Purchase, and Philosophy at UMass/Amherst and UCLA. She attended UCLA Law School and is a lawyer, homemaker, and soccer mom in Davis, California.

Marc Hauser was born in Cambridge, Massachusetts. He completed his undergraduate degree in Animal Behavior at Bucknell University and his Ph.D. in Behavioral Biology at UCLA. Post-doctoral work was performed at the University of Michigan, Rockefeller and the University of California at Davis. Currently he is Associate Professor of Anthropology and Psychology at Harvard University. He is interested in the evolution of cognitive abilities and language, and studying monkeys and apes in the wild and in captivity.

John Heminway

In 1994 and 1995, John Heminway worked at Disney Cruise Lines, helping fashion a storyline and ship design to distinguish the two 85,000-ton Disney ships and their on-board experiences from all others in this booming industry. In April 1995, Heminway conducted Disney executives on an East African safari, in preparation for the launch of Disney's "Animal Kingdom" in Orlando. Heminway also spearheaded the East African premiere of Disney's *The Lion King.* He has served as program consultant for the Disney Institute, and for the Disney Vacation Club. Since 1987 John Heminway has been the Chairman of the African Wildlife Foundation. He works out of homes in New York and Montana.

A. Magdalena Hurtado

A. Magdalena Hurtado was born in Caracas, Venezuela. She completed her undergraduate degree at SUNY Purchase in 1980. With a *Gran Mariscal de Ayacucho* award to Venezuelan nationals to fulfill requirements for the Ph.D. she went to Columbia University and the University of Utah. Her dissertation research focused on women's foraging strategies among Ache hunter-gatherers of Eastern Paraguay. Subsequent work focused on similar problems in diverse lowland South America indigenous groups. Since 1988, Hurtado has been involved in a longterm project designed to study mothering among American ethnic groups. Hurtado is now Assistant Professor in the Human Evolutionary Ecology Program, Department of Anthropology, at the University of New Mexico where she teaches evolutionary medicine.

Andrew Grieser Johns

Andrew Grieser Johns has worked in Amazonia—where this story is located, Southeast Asia, and East Africa. His main professional interest has been the effects of timber harvesting on the wildlife of tropical forests and the design of forest management operations that limit ecological and environmental damage. He is currently a full-time houseperson, living in Cambridge, England.

Kate Kopischke

Kate Kopischke is a free-lance writer and graphic designer in Albuquerque, New Mexico. She has lived and worked with two groups of Native South Americans in Peru— the Piro and the Machiguenga—and with a village of mixed ethnicity in Botswana's Okavango Delta.

Phyllis Lee

Phyllis Lee was born in California, and first experienced the joys of field work on baboons in Tanzania. She then migrated to Cambridge (England), and did her Ph.D. research on vervet monkeys in Kenya. For her subsequent research, she shifted up the body size scale to elephants. She is currently a University Lecturer in biological anthropology at the University of Cambridge, England.

Herbert H. T. Prins

Herbert H. T. Prins was born in the Netherlands into a family of four brothers and four sisters. His father, an anthropologist, did most of his fieldwork in East Africa, so as a child he knew he wanted to work there. He received his training as an ecologist at the University of Groningen in the Netherlands as a student of Rudi Drent and later at Cambridge University (Darwin College). His supervisor, Tim Clutton-Brock, suggested a Ph.D. study on the African buffalo in Manyara, Tanzania. He then worked in a World Bank project on National Park buffer zone and research management in Indonesia, and as research-fellow of the Royal Netherlands Academy he continued his buffalo work in Manyara. In 1993 he became Full Professor in tropical nature conservation at Wageningen Agricultural University (the Netherlands), and he continues working the summer terms in Tanzania.

James Serpell

After completing his fieldwork in Australia and Indonesia, James Serpell returned to Britain to complete his Ph.D. at the University of Liverpool. (His long-suffering lorikeets eventually joined the collection of the North of England Zoological Society at Chester Zoo.) Continuing his commitment to eccentric research activities, he then moved on to study the behavior of domestic dogs and cats and their relationships with people at the Sub-department of Animal Behavior in Cambridge. He is now Director of the Companion Animal Research Group at Cambridge Veterinary School. James Serpell is the author of *In the Company of Animals* (Basil Blackwell, 1986).

Kelly Stewart

Kelly Stewart first went to the Karisoke Research Center in Rwanda in 1973 after graduating from Stanford University with a B.A. in Anthropology. She was a research assistant to Dian Fossey and then conducted a study on the social development of immature mountain gorillas, receiving a Ph.D. in Zoology from Cambridge University in 1981. She returned to Rwanda from 1981-1983 as Co-Director of the Karisoke Research Center with her husband, Sandy Harcourt, whom she had met in Rwanda amidst the gorillas. Kelly Stewart is currently a research associate in the Anthropology Department, University of California at Davis. She edits the yearly *Gorilla Conservation News* and is working on a book about the Nigerian expeditions. She also serves as a scientific advisor to the Dian Fossey Gorilla Fund that supports the Karisoke Research Center, still operating twenty-seven years after being established by Dian Fossey.

John Symington

John Symington lives in Washington, DC with his wife and three children. He is a physician specializing in infectious diseases and has done research on AIDS. He is currently in private practice. He sings with a small a capella group and enjoys bird watching.

Margaret Symington

After completing her study of spider monkeys and receiving her doctorate in biology from Princeton in 1987, Margaret went on to focus her efforts on the conservation of tropical forests and biodiversity in Latin America. She spent two years as an environmental adviser for the United States Agency for International Development (USAID), and now coordinates conservation activities in Latin America for the Biodiversity

Support Program, a USAID-funded consortium of the World Wildlife Fund, The Nature Conservancy, and World Resources Institute. She and her husband John, a physician, live in Washington, DC.

Monica Udvardy and Thomas Hakansson

Monica Udvardy has a Ph.D. in Cultural Anthropology from Uppsala University and is an Assistant Professor at University of Kentucky. She conducted a total of seventeen months of fieldwork on women and the lifecourse among the Giriama of Kenya between 1983 and 1986. Her interests and publications concern the cultural construction of gender, medical anthropology, and gender issues in development.

Thomas Hakansson has a Ph.D. in Cultural Anthropology from Uppsala University and is an adjunct Associate Professor at Uppsala University. Among his publications are *Bridewealth, Women and Land: Social Change among the Gusii of Kenya,* and numerous articles on marriage and economic processes in eastern Africa. His research interests are Economic and Political Anthropology and he is currently engaged in an ethnohistorical project on political centralization and regional interaction in pre-colonial northeastern Tanzania.

Pieter van den Hombergh

Pieter van den Hombergh studied medicine at the University of Nijmegen, The Netherlands. From 1980 until 1984 he worked as a tropical doctor in St. Joseph's Hospital in Kilgoris, Kenya. After his return he became a Family Doctor in a health center in Almere, a new town in an area retrieved from the sea. He also works part-time as a health researcher in Nijmegen developing methods for assessing practice management.

Truman P. Young

Truman P. Young was born and raised in Colorado, and has lived much of his life in and around mountains. He carries out ecological research in East Africa, Panama, and North America, studying plant populations and communities and their interaction with wildlife and humans. After years as a tropical derelict and later at Fordham University, he recently joined the faculty of the University of California at Davis. He and his wife, behavioral ecologist Lynne Isbell, have a two-year-old son, Peter.